To Jack

Good Luck with the
color and conformation of the
babies of your grey mare.

Merry Xmas "96"

Love Jan

HORSE COLOR

Horse Color

by D. Phillip Sponenberg
and Bonnie V. Beaver

Library of Congress Cataloging in Publication Data

Sponenberg, D. Phillip (Dan Phillip), 1953–
Horse color.

Bibliography: p.
Includes index.
1. Horses—Color. 2. Horses—Breeding.
3. Horses—Genetics. I. Beaver, Bonnie V. G.,
1944– . II. Title.
SF279.S67 1983 636.1 83-45102

ISBN 0-914327-46-1

Reprinted in 1994 in Colombia, South America by Carvajal S.A.

To Torsten, Henry, Anne, and the giant forest horses
D.P.S.

To Mom, Dad, Larry, Bud, and Stubby
B.V.B.

Contents

Preface

The project started as a slide program, but the idea for a book became pressing when we realized that no complete guide to horse coat colors had been published in English. Good books on color do exist in Dutch and Spanish, but these are usually not available to English-speaking people. Neither English terms nor Spanish terms alone express all of the details of horse color so it seemed best to combine them in a guide that would give a complete explanation of horse color terminology.

We hope that one result of this guide is the eventual recording of color on registered horses in an accurate, complete manner that will allow definitive genetic research to be accomplished in the future. The present status of the genetics of horse color leaves much unexplained, and the frequency with which the present rules do not hold true indicates that the present explanations are inadequate.

The book's photographs are not all of beautiful horses in a show stance. Some of them are of very poor examples of livestock, but all, we hope, clearly and accurately show the colors discussed. The mixture of the good standard livestock photographs with the more casual ones may even help keep the reader interested. The book was fun to put together, and we hope it will be enjoyable to read.

This book would have been impossible without the contributions of a great many people. Discussions of horse color with Henry and Anne Harper, breeders of Brabant and American Belgians, were most enjoyable and sharpened a lot of the fuzzy areas. Their knowledge of horses and horse color is phenomenal, and seeing their 2,500-pound stallion float over the ground in a perfect trot proves that the classical approach to horse breeding and evaluation is still quite successful. Buddy and Leana Rideout, breeders of American Indian Horses, first introduced Phil Sponenberg to the Spanish horse and the array of colors in the breed. Their assistance and friendship have been valued highly over the years. They are also proving that the Spanish horse and its capabilities are by no means diminished in the present survivors of the breed. Marye Ann Thompson, a breeder of Spanish Mustangs, has helped a lot in general discussions of color as well as in the study of the genetics of color. She was especially diligent in calling attention to well-documented exceptions to the theories of inheritance of coat color in horses.

J. K. Wiersema, the author of an excellent Dutch book on horse color, has helped immensely with ideas on genetics and the classification of color. His informative letters of the last few years have been a real treat. Gilbert Jones, a breeder of Southwest Spanish Mustangs, is yet another who understands the Spanish horse. His work of preserving the Spanish horse in an array of colors has been very helpful to this book. Many of the photographs of Spanish horses in the text are of horses not owned by him but from his bloodlines, and so the credit lines do not accurately reflect the contribution he has made. Jim and Sharon Babbit, breeders of Spanish Mustangs, have also been helpful.

Discussion of genetics with F. B. Hutt has been extremely helpful. He would say that all the genetics we present are "iffy" because no numbers and statistical tests are given. We wholeheartedly concur, but we feel that the inheritance pattern proposed here will stand the tests of time, and it does explain hitherto unexplained colors. L. D. VanVleck has likewise been of great assistance in discussions of genetics as well as of color in general. H. F. Hintz provided the lilac dun pony to photograph, and because of the rarity of that color, it deserves mention here. Larry and Joyce Slusser, breeders of Quarter Horses, and Alda Buresh, a Saddlebred breeder, were helpful in guiding us to the last horses photographed; those represented some of the more difficult colors to obtain. George Hatley of the Appaloosa Horse Club helped to clear up some of the definitions used by that group.

Many other people, including Nanci Falley, Kim Kingsley, Doug Gregg, Sue Quick, Debbie Dinnan, Linda Till, Gene Palmer, Bud Alderson, and Joyce Smith, have also been of tremendous help. Their contributions are also acknowledged in the text. The manuscript was typed by Gayle Anderson, Barbara Barnett, Dawn Bender, Diane Bolling, and Joyce Reyna. Details of duplicating photographs and general technical support were supplied by Jerry Baber. Sharon Ashby contributed the beautiful artwork. Sandy Berry and Joyce Slusser helped with details of various sorts. A big thanks is also given to owners whose names we do not know and to all who held the horses for the photographs.

Horse Color

Introduction

Horses come in a wide range of colors that can be subtle and confusing. The names used to describe them have become equally confusing. The most extensive and exact vocabularies describing colors and patterns of white on horses were used in the American West, Argentina, and Mexico. Much of the detail used in these descriptions arose because nearly all possible horse colors and patterns occur in the Spanish horse that populated those regions. These colors and patterns can still be seen in breeds that were derived from those Spanish horses: American Quarter Horse, Paint, Pinto, Appaloosa, Criollo, and North American Spanish Horse (including such registries as Spanish Mustang, American Indian Horse, Spanish Barb, and Southwest Spanish Mustang). In recent years some of this detail of description has been lost to horsemen or has come to be thought of as superfluous.

This book attempts to include detail that has an important biologic basis (largely but not solely mediated by genes) and to omit most of the detail that arises from environmental differences in nutrition or season of the year. The terminology that developed in the American West is used throughout the text but is expanded to include the names of colors used in other English-speaking areas when appropriate. Spanish is used where necessary to supplement the English because some of the names have no English equivalent. No single language has names for every variation in color and pattern, but a combination of English and Spanish names is quite complete.

In this book, *color* refers to nonwhite areas on the horse, and *pattern* refers to white hair and its arrangement on the animal. This approach greatly simplifies the understanding of the colors of the horse as well as the patterns of white. A special vocabulary of names has developed over the centuries to describe the specific combinations of colors and shades and their locations on the horse (Fig. 1). These special horse color terms appear in italics throughout the text in order to distinguish them from words describing color per se. For example, "*brown*"

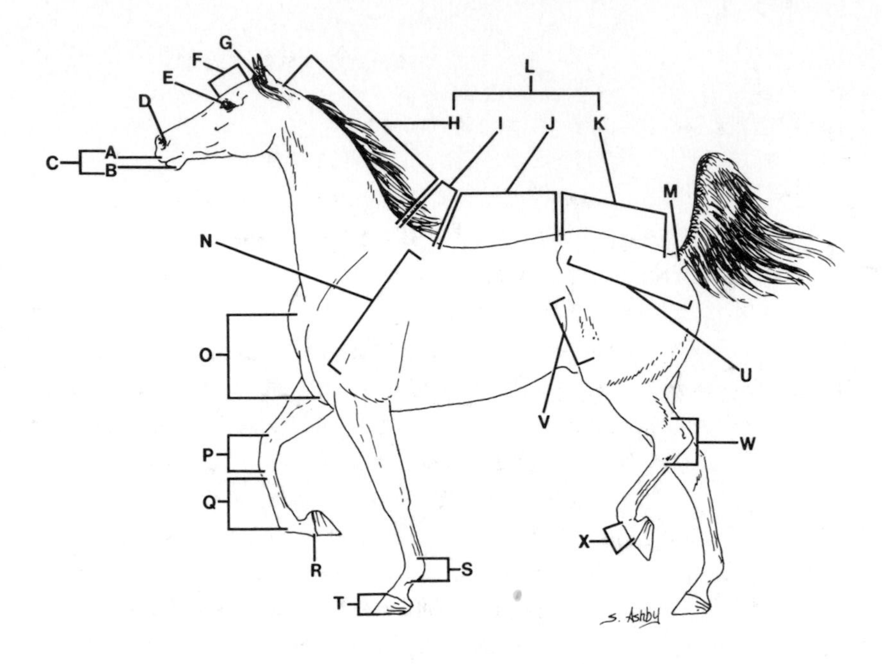

1. Horse conformation: *A*, upper lip; *B*, lower lip; *C*, muzzle; *D*, nostril; *E* eye; *F*, forehead; *G*, forelock; *H*, neck; *I*, withers; *J*, back; *K*, croup; *L*, topline; *M*, base of tail; *N*, shoulder; *O*, chest; *P*, knee; *Q*, cannon; *R*, coronary band; *S*, fetlock; *T*, foot; *U*, hip; *V*, flank; *W*, hock; *X*, pastern

is the specific horse color term, while "brown" refers to the color with no particular reference to horses. In this naming scheme, each animal can be described by a single horse color term such as *black*, *chestnut*, *grullo*, *palomino*, or *sorrel*. Even horses having more than one color or color shade in their hair coat can be described in this simple manner, because "horse color" is the sum of "body color" and "point color." The term *bay*, for example, denotes a red body with black mane, tail, and lower legs. Although two colors are present (black and red), one horse color name is used to indicate the combination.

In the recent past, horses displaying patterns of white were described only by the pattern present (*paint*, *appaloosa*), while references to the color were omitted. A more accurate approach is to use the single horse color name and then to add the names describing all patterns of white present. Examples are *bay tobiano* and *buckskin roan.* The captions of the illustrations throughout the text use this system of a basic horse color name followed by the white pattern name. White or mostly white horses will confound this approach, but such horses are rare and can be described simply as *white.*

Horse Colors

An easy way to classify horse color is first to consider the color of the "points" of the horse (Fig. 2). The points are the mane, tail, ear rims, and lower legs. All horses are divided into two

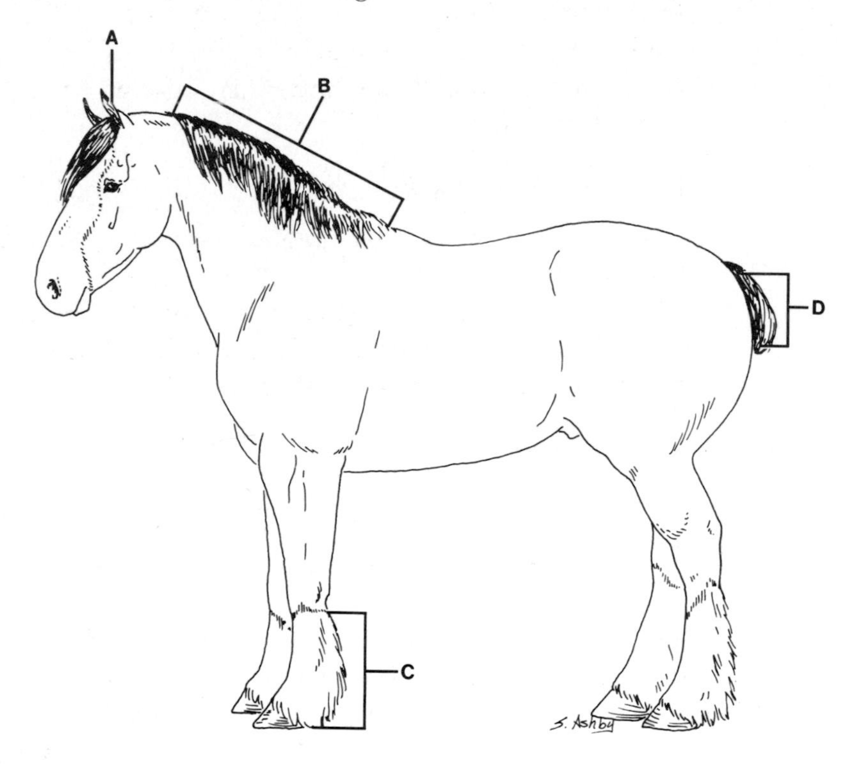

2. Color points of the horse: *A*, ear rims; *B*, mane; *C*, lower legs; *D*, tail

groups on the basis of the color of the points: horses with black points and horses with nonblack points. The nonblack points span a spectrum from brown to red to cream.

Black and nonblack points are usually visually distinct, as in Fig. 3. Exceptions include some black manes and tails that fade, appearing nonblack (Fig. 19), and the manes and tails on some horse colors with nonblack points that are so dark that they look almost black (Fig. 36). Because of possible confusion concerning mane and tail color, the lower leg is frequently a more accurate indicator of point color. The point color usually is obvious, and deciding whether the horse has black or nonblack points is generally easy. On most horses with nonblack points the color is lighter in that area. Confusion can arise on this subject because some dark brown points are fairly uniform even to the hoof and can be mistaken for black (Fig. 62). Some black points are not extensive, and thus lighter areas may extend down to the lower leg and may therefore have the appearance of nonblack points (Fig. 21). Horses of identical body color can have different point colors, so differences in point color are very important in choosing the accurate horse color term. Fig. 3 shows black and nonblack points on horses with identical body color. Many horses have white on one or more lower legs (Fig. 15), but since white is not a color (it is the absence of color), it must be ignored when the color of the points is being considered.

The black-points group and the nonblack-points group are the largest natural divisions of horse colors, but each consists of smaller subgroups of colors that are related.

Horse Colors with Black Points

The various groups of horse colors with black points are *black* (body black), *brown* (body brown), *bay* (body red), *grullo* (body slate), and *buckskin* (body yellow). Occasionally borderline body colors will occur and will present problems in naming the color. Generally, however, the decision is obvious.

Black

Black horses are rare in most breeds except the Percheron; the Fell Pony; the North American Spanish Horse, wherein *blacks* are relatively common; and the Friesian, which is always *black. Black* horses have black points and black bodies with no obvious hairs of other colors, although white markings on the face and legs can occur. If a detailed search is needed to turn up lighter hairs on a *black* horse, it is still accurate enough to consider it *black.* Figs. 4 and 5 are typical of most *black* horses. The body is slightly lighter than the points. After prolonged exposure to the sun, many *blacks* develop a rusty color, but this turns to black again when the horse sheds, as is happening in Fig. 6. Occasionally *black* horses have faded manes and tails, as in Fig. 7. *Jet black,* as illustrated in Fig. 8, has a body color that is the same pure black as the point color, even in strong sunlight. The color in Fig. 9 is called *smoky black* and looks more off-black than truly black. When compared with true *blacks; smoky blacks* have a body color that is quite a bit lighter than the points.

Brown

Brown consists of fairly uniform shades of body color that occur between black and red (*bay*). *Brown* is used to describe those horses with black points whose bodies are a shade of brown with no red in it. Fig. 10 illustrates a dark *brown,* which can be very close to *black* or *smoky black* but generally has a different, browner shade than that of even the faded *blacks.* The middle shade of *brown* is a flat, dull color, as in Fig. 11. *Light brown,* as in Fig. 12, can be confused with *bay* and is sometimes classified with it; however, it lacks the red shade that gives the *bay* body color its characteristic brilliance.

Seal brown is a special type of *brown* that is close to *black. Seal brown* can be distinguished from *black* only because lighter brown or yellow areas occur on the muzzle, over the eyes, on the flanks, and on the insides of the legs. Fig. 13 is a dark *seal brown* on which the lighter areas are inconspicuous. On light *seal browns,* such as Fig. 14, the light areas are quite noticeable. Fig. 15 illustrates the difference between the *seal brown* horse and the *jet black* one standing next to it. This picture also serves to demonstrate that high white leg markings can obscure lower leg color. *Seal brown* may represent one of the ancestral colors of the horse, since it is seen in the Exmoor Pony, which is a primitive breed.

Bay

Bay describes horses with black points and with bodies that are colored some shade of red. *Bay* is a common color and can be divided into several shades. At one extreme the body color is nearly yellow, and at the other extreme it is mixed with black and the horse can resemble a *brown*, *seal brown*, or *black*. All *bays* have red in the coat, which usually gives the color a certain brilliance and sheen.

Figs. 16 and 17 illustrate the most common *bay*—*red bay* (*cherry bay*). The body is a clear shade of red with little variation in intensity. *Mahogany bays*, such as Fig. 18, are sometimes referred to as *dark bay* and result from black being mixed into the red body coat. The black is usually more abundant dorsally (along the topline), especially over the croup and withers. The black can be in the form of individual black hairs or black tips on the red hairs. In the darkest extreme, *mahogany bays* can be confused with *seal browns*, and indeed the horse in Fig. 13 looks *bay* instead of *seal brown* in some years. Another *mahogany bay*, in Fig. 19, illustrates that on some horses with black points, the mane and tail can fade to brown. *Blood bay*, seen in Fig. 20, is another dark shade of bay, but instead of being a mixture of red and black as in *mahogany bays*, the body color is a pure shade of dark purplish red with little or no black intermingled. The lightest *bay* shade is *sandy bay* (also called *honey bay* or *mealy bay*), seen in Fig. 21. In *sandy bay* the body is a light red that approaches yellow. Fig. 21 also shows a horse with black points that do not extend very far up the lower leg, demonstrating the importance of looking closely at the color near the hoof of such horses.

Dun

While some horsemen use *dun* to indicate yellow horses with black points, the term is frequently used in a general sense to describe all of the lighter horse colors, some of which do not have black points. The terminology of the American West, Argentina, and Mexico used *dun* (*bayo* or *gateado* in Spanish) for all light colors and then modified the word *dun* by names that indicated the body color and point color. These colors are common in Quarter Horses, the North American Spanish Horse, and the Criollo as well as in some European breeds such as the Norwegian Fjord Horse and the Icelandic Horse. People familiar with these breeds use a more detailed nomenclature than do people from England (where the light colors are rarer), who are likely to use *dun*, unmodified, to describe the whole group of light shades. As a group, *duns* tend to have points darker than the body color. The base of the tail and sides of the mane are frequently lighter than the remaining tail and mane hairs, as can be seen in Fig. 28.

The *grullo* and *buckskin* horse colors are both types of *duns* having black points. The other types of *duns* will be discussed in the section of horse colors having nonblack points.

Grullo. *Grullo* (pronounced GREW-yo) is the Spanish name for the sandhill crane, a slate-colored bird. This term is used by Western riders when referring to a blue slate–colored horse with black points and a dark or black head. *Mouse dun* and *blue dun* are English terms describing the same color but are used more by English riders or when describing breeds from Europe. *Grullo* horses almost always have primitive marks (withers stripe, dorsal stripe, and stripes over the knees and hocks).

Slate grullo is the usual *grullo* color and is seen in Figs. 22 and 23. The body is a slate or tan color with only minor differences in shade, and the head is quite dark. Fig. 24 shows a *grullo* in which the slate-colored body hairs are mixed with black, resulting in a darker body color, especially dorsally. This color is referred to as *lobo dun* (*lobo* is Spanish for "wolf"). The horse in Fig. 25 has black points, a dark head, but a yellower body color than that of most other *grullos*, a combination referred to as *olive grullo* (or *olive dun*) because it can approach the color of an unripe olive. When black is mixed with another body color, the horse's color is described as *smutty* or *dark*. If the olive body color is mixed with black, as in Fig. 26, the result is called a *smutty olive grullo*. *Silver grullo*, as in Fig. 27, is the lightest shade of *grullo*. The body is a cream color, and the points and head are a slate blue of varying intensity instead of the black of most *grullos*. *Silver grullos* have blue eyes.

Some variations in the body color of *grullos* can occur as a result of environmental changes. Most *slate grullos* tend to fade to *olive grullo* in strong sunlight; however, some *olive grullos* are always olive and are not faded *slate grullos*. Thus, the two colors are distinct. *Grullo* of varying shades occurs as an ancestral color in the Tarpan, a reconstructed wild horse of Poland.

Buckskin. Confusion concerning the *buckskin* horse color designation has resulted from the loss over the years of the detailed terminology of the early American West. This group of *duns* consists of horses that are a shade of yellow with black points. The head is the same color as the body or only slightly darker. Some horses in the *buckskin* group have primitive marks; others do not.

Some horsemen separate the horses with yellow bodies, black points, and primitive marks and call them "*dun*," reserving "*buckskin*" for those without the marks. Others completely reverse this, using "*dun*" for the unmarked ones. This has been a source of great confusion. In the past, and in the scheme presented here, the general group of yellow-coated horses having black points is called *bucksin*. In addition, the term *buckskin* also specifically designates the unmarked members of this group, as in Fig. 28. When primitive marks are present, as in Fig. 29, the result is called a *zebra dun*; this is the color most frequently referred to simply as *dun*, but *zebra dun* allows for greater detail and less confusion. If the *buckskins* or *zebra duns* have a brownish cast to their yellow coats, they are called *dusty buckskins* or *dusty duns* (Fig. 30), and they appear similar to *olive grullos* without the dark head. Members of the *buckskin* group having black hairs mixed into the body coat are called *dark (smutty) buckskins* (Fig. 31). The *coyote dun* is the primitive-marked equivalent of the *dark buckskin*. Fig. 32 shows a *silver buckskin*, which has a creamy body color and no primitive marks. *Silver dun* is the same color with primitive marks, as in Fig. 33.

Olive grullos are sometimes included in the *buckskin* group, but their very dark heads place them more accurately in the *grullo* group. *Zebra dun* occurs as an ancestral color in the Norwegian Fjord Horse and the Highland Pony.

Horse Colors with Nonblack Points

The various subdivisions of horse colors with nonblack points are *chestnut/sorrel* (body red), *red dun* (body very light red), *yellow dun/palomino* (body yellow), *silver dapple* (body sepia brown), and *cremello/perlino* (body off-white). Overlap and confusion about terminology are greater in this group than among the horse colors with black points; however, a close look at the groups reveals fairly good divisions.

The color of the points in this group deserves special mention, being brown, red, or flaxen (cream). Usually the lighter the mane and tail, the lighter the lower leg. The mane and tail are usually the same color, although they can be various combinations of red and flaxen, such as flaxen mane with red tail or red mane with flaxen tail. Horses with either dark brown or very light (off-white) manes or tails usually have the same color on both mane and tail. The red and darker flaxen manes and tails can change color with age, but do so only rarely. Some manes are a mixture of red, flaxen, and brown hairs, as are some tails.

Chestnut/Sorrel

In English-speaking countries this large horse color group is usually separated by body color into several divisions, including *liver chestnut* for darkest red, *chestnut* for medium red, and *sorrel* for light red. In Argentina (Spanish) the breakdown is by point color: *tostado* for brown points, *alazán* for red or dark flaxen points, and *ruano* for the light flaxen points. Combining the two languages allows a more accurate description of the individual horses.

Distinguishing between *sorrels* and *chestnuts* depends largely on the breed under consideration. In some breeds, such as the Thoroughbred, Arabian, Morgan, and Suffolk, all shades of red are classified as *chestnut*. Although in the Suffolk breed all are designated as *chestnut*, the color is still subdivided into seven shades: *yellow*, *light*, *copper*, *gold*, *red*, *dark*, and *liver*.

The usual approach with draft horses, particularly the American Belgian, is to go by the number of easily distinguished shades of red or lighter color on the horse—two or fewer being a *chestnut* and three or more a *sorrel.* Most of the horses with three or more shades also have very light points or conspicuous light areas on the flanks, on the insides of the legs, over the eyes, and on the muzzle. By these draft-horse criteria, the reds of the Thoroughbred, Arabian, Morgan, and Suffolk breeds would indeed be *chestnut,* not *sorrel.*

In the American Quarter Horse breed the light, clear reds are considered *sorrel,* regardless of the number of shades present. The medium reds are *chestnut,* and the very dark reds are *liver chestnut.* In any system adopted, the boundaries for the *sorrels, chestnuts,* and *liver chestnuts* are very subtle, and some horses change shades during their lifetimes. Some horsemen have decided simply to call the whole group "red" to avoid confusion, but that approach does not reflect the vocabulary generally used. The approach taken here is most like that of the Quarter Horse color designations, because dividing the red shades into three basic groups, with the addition of the Argentine Spanish terms for the color of the points, is probably adequate for most uses.

The *chocolate chestnut* is a rare color in this group. As illustrated in Fig. 34, the body color is uniform chocolate, and the points are brown. The actual shade of the body color can vary from dark to medium brown. Fig. 35 shows how similar *liver chestnut tostado* can be to *chocolate chestnut,* except that the lower leg is lighter than the body.

Liver chestnut is the darkest of the red shades. It is usually a mixture of red and black but can be a uniform shade. Fig. 36 illustrates a *liver chestnut tostado.* The color of the legs does get lighter toward the hoof of this individual. Figs. 37 and 38 are *liver chestnut alazáns.* Body color on these is a mixture of dark red and black, which cannot be readily appreciated from pictures. Fig. 39, a *liver chestnut ruano,* illustrates how light the mane can become. Another *liver chestnut ruano,* Fig. 40, shows points typical of this horse color.

A *chestnut tostado* is shown in Fig. 41, and a darker *chestnut tostado* in Fig. 42. The *chestnut alazáns* in Figs. 43 and 44 demonstrate the mixture of dark and red in the coat. *Chestnut ruano* is pictured in Figs. 45 and 46.

Sorrels have clear red coats which combine with the various nonblack point colors. A *sorrel tostado* is pictured in Fig. 47, and a *sorrel alazán* is in Fig. 48. Fig. 49 shows *sorrel alazán* with a light mane but dark tail. Fig. 50 is a *sorrel alazán* Belgian with light areas on the body. The *sorrel alazán* horses in Fig. 51 have manes and tails that are lighter than the body, but not the near white of the *ruano*. The flaxen mane and tail of the *ruano* go along with its lighter lower legs and can make it look almost like a *palomino*. Fig. 52 illustrates the lightest *sorrel* shade, the *blond sorrel ruano*. The body is a light sandy red with pale areas over the eyes, on the muzzle, on the flanks, and on the insides of the legs. The lower legs are quite pale also. This color is most common on the American Belgian Draft Horse.

Dun with Nonblack Points

Dun as a color group is confusing when terminology of the early American West is ignored, so the reader should refer back to the general comments on *dun* horse colors with black points, above. In the past, *duns* with nonblack points were split into *red dun* and *yellow dun* (including *claybank* and *palomino*). *Red dun* includes the specific colors of *muddy dun*, *red dun*, *orange dun*, and *apricot dun*, while *yellow dun* includes *yellow dun*, *claybank dun*, and *palomino*. The term *claybank dun* has more recently been used to describe all *red* and *yellow duns* except *palomino*.

An extremely rare horse color that does not fit neatly into the *red dun/yellow dun* groups is the *lilac dun* (*dove dun*). These horses have chocolate brown points and a lilac or dove-colored body, as in Figs. 53 and 54. The skin is a light brown or pink, and the eyes are usually amber.

Red Dun. The variations of this group have very washed out red or yellowish red bodies; brown, red, or flaxen points; and, usually, primitive marks. Although these horses colors are sometimes called "claybank duns," that color designation is also used for a specific shade of the *yellow dun* group, so the term *red dun* is preferred for this group. The darkest shade in the group is *muddy dun* (Fig. 55). The body is a light brownish red or brown yellow, and the points are chocolate brown (Fig. 56). Because the head is usually chocolate brown, too, the color resembles *grullo* except that the black and slate have been replaced by brown and pale brownish red. *Red dun,* as in Figs. 57 and 58, lacks the brown head of the *muddy dun.* The body is a light red, and the mane, tail, and lower legs are usually a darker red. *Orange dun,* as in Fig. 59, has a lighter body color. The lightest color in this group, *apricot dun,* is seen in Figs. 60 and 61. The same horse is shown in winter and summer to demonstrate the differences that can occur in any color with changes in season. The body is light red, tending to a yellow shade. The mane and tail can be pale red, brown, or flaxen.

Yellow Dun. This color term is sometimes used to describe *buckskins,* but it is more accurately applied to horses that have yellow bodies and nonblack points. Differences between some of these body color shades and the shades of the *red dun* group can be slight. *Yellow duns* are yellow with brown points and resemble *buckskins* or *zebra duns* except that the points are brown instead of black. Fig. 62 illustrates the difference between *yellow dun* and *zebra dun.* The *claybank dun,* as in Figs. 63, 64, and 65, is essentially the shade between *apricot dun* and *palomino.* The difference between *claybank dun* and *apricot dun* can be slight, with the *claybank dun's* body color being paler or yellower. Some resemble *palominos* but lack the very light mane and tail of the *palomino.*

Palomino. The *palomino* belongs to the *yellow dun* group, but the recent trend has been to consider it as a separate group from the other duns. *Palominos* are a shade of yellow (or gold) with light mane and tail (Figs. 66 and 67). The *palomino* in Fig. 67 is dappled, which is common on *palominos.* Occasionally *palominos* and other *duns* have pink skin, as in Fig. 68. *Smutty (sooty) palominos* have a mixture of yellow and black over the body, as in Fig. 69, and can be as dark as some *chestnuts* except that they lack the red tint that *chestnuts* have. In Fig. 70 the black has also mixed into the tail. The term *isabella* describes the very light cream–colored *palominos* having nonblue eyes, as in Fig. 71. *Isabella* is the term currently being used in Europe for all *palominos.* In the United States, *palomino* registries restrict the permissible color to only certain shades of yellow, so *isabella* takes on a restricted meaning.

Silver Dapple

Silver dapple is a special horse color which may be confused with *chestnut*. Although in the United States *silver dapple* occurs only in Shetland Ponies, it can be seen in a wide variety of breeds in Europe. As Fig. 72 shows, the body color is sepia brown with light dapples, and the points are flaxen or nearly white. The dapples may be lacking or subdued in some individuals (Fig. 73), but the points have a distinctive appearance that usually identifies the animal as a *silver dapple*.

Cremello/Perlino

Two colors with nonblack points that are nearly white are the *cremello* and the *perlino*. The *cremello* (Fig. 74) has a nearly white, cream-colored coat and points and blue eyes. The *perlino* has a similar coat and eyes but has slightly red or blue points. The *perlino* in Fig. 75 illustrates the slightly red mane. Both *cremello* and *perlino* can be easily mistaken for white, but close inspection reveals that they are cream-colored instead. A color difference at the border of a white facial or leg marking is the most reliable way to detect these off-white body colors.

Miscellaneous Effects on Horse Color

Dappling

Dappling is a network of darker areas imposed over lighter areas and can be present on horses of any color (Figs. 67, 76, 77, 94, 95, and 187). The effect is the result of some fundamental differences in the hair of various areas of the skin, so *dapples* can be seen even on black horses (Fig. 77). *Dappling* is generally most noticeable when horses change hair coats in the spring and fall, but it is present throughout the year on *silver dapples* and many *greys*.

Pangaré

Pangaré (pronounced pahn-gah-RAY; meaning "mealy muzzle") is an arrangement of light areas that can be superimposed over any color. These lighter areas are over the muzzle, over the eye, inside the legs, and in the flanks. Since this pattern is rarely mentioned in English-speaking countries, the Spanish term has been retained. When *pangaré* is imposed over *black*, the result is *seal brown*; when over *sorrel*, the result can be *blond sorrel*. This is the factor that distinguishes *sorrel* from *chestnut* in the color-naming scheme of the Belgian Draft Horse. On most horse colors the effect does not change the descriptive name, as in the *mahogany bay* in Fig. 78, which is also a *pangaré*.

Primitive Marks

Primitive marks include a stripe down the back (*list* or *eel stripe*), a stripe over the withers, and stripes over the knees and hocks (*zebra* or *tiger stripes*). The line down the back may be black (as in Fig. 79), brown, red (as in Fig. 80), or gold, and it can occur on any coat color. Horses possessing it are called *linebacks*. In Fig. 81 the dark dorsal stripe continues as a darker portion of the tail, which is common in several colors with dorsal stripes. Stripes on the withers and neck are illustrated in Figs. 80 and 82. *Zebra stripes* on the legs, another primitive mark, are illustrated in Figs. 83 and 84. Any of these primitive markings can be present in any combination and on any horse color, as seen on the *bay* in Fig. 82. They are, however, most commonly observed on the *dun* colors, since darker colors obscure their presence. Very rarely a horse will be extensively striped, almost to the extent of resembling a zebra. Such horses are known to occur in Siberia, Scandinavia, and Argentina.

Foal Colors

The colors of foals are usually not easy to categorize. This can be a source of confusion if classification of an animal's color for registration occurs before the foal coat is shed, since the early coat is usually lighter than the adult coat, as is obvious in Fig. 85. Many *black* horses, for example, are born with a slate body and point color, which then sheds to black (Fig. 149). Black points on foals can similarly be light and resemble nonblack points, as in Fig. 86, a *linebacked bay* foal. The foal in Fig.

87 will gradually shed the reddish foal coat to reveal the *grullo* color of the adult coat. Other *grullo* foals are an ash color even on the head, but the dark head typical of *grullos* is revealed when the foal coat is shed (Fig. 88). Therefore, foal color is not necessarily a good indication of adult color.

Eye Color

Eye color is variable in horses. Most have dark brown eyes (Fig. 89). Amber (*hazel*) eyes occur (Fig. 90) and are especially common on the *dun* colors. Some *duns* have grey eyes. Blue eyes frequently occur if the skin around the eye is unpigmented (pink) (Fig. 150), but a number of horses with blue eyes have normally pigmented skin (Fig. 91). Conversely, some eyes in an unpigmented field are brown. Blue eyes, also called *wall eyes*, *glass eyes*, or *china eyes*, are present on all *perlinos*, *cremellos*, and *silver grullos* (Figs. 75 and 92).

Seasonal Color Changes

Some horses change color from season to season or occasionally from year to year. Such changes are not marked and usually affect the body and not the points. The color of horses that do change will usually remain within the same color group; for example, a *red bay* may change to *mahogany bay*. Occasionally, though, a change will occur to a closely related group, as from *chestnut* to *sorrel* or *mahogany bay* to *seal brown*. Most horses, fortunately, do not undergo such changes.

Patterns of White

Several patterns of white hair occur on horses. Each horse should be described by the appropriate horse color term followed by appropriate descriptions of the patterns of white present on the animal. While some patterns of white are mixtures of white hairs with colored hairs, others are patches of solid white. Some breeds have been bred and selected on the basis of certain of these patterns. For example, asymmetrical white patches are typical of Paints and Pintos, while symmetrical patterns of white patches are typical of Appaloosas and also occur in other breeds such as the North American Spanish, Knapstrub, Noriker, and Pony of the Americas.

Patterns of Individual White Hairs

"Roan" is the general term used to describe mixtures of individual white and colored hairs in animals. However, with horses this term creates confusion, since it is used both in this general sense and for a specific pattern. The specific pattern called *roan* in horses is a fairly uniform mixture of colored and white hairs on the body, with the head and points exhibiting only colored hairs. *Roan* is a nonprogressive pattern; that is, the horse is born *roan*, or sheds to *roan* after the foal coat, and changes little from then on. *Roans* are very rarely dappled. *Grey* horses also fit the general classification of "roan." But in contrast to the specific *roan* pattern, *greys* are born solid-colored and then become progressively lighter as they age. Consequently, most old *greys* are white with dark skin. *Greying* affects the points and head, as well as the body, and *greys* are frequently dappled. Some registries, most notably the Thoroughbred, reserve the term *grey* for black or dark horses undergoing the greying process, because these combinations truly look grey. They then use the term *roan* to describe colors undergoing the greying process in horses that do not look grey (lighter *bays*, *chestnuts*, and *sorrels*). Because the basic patterns of *roan* and *grey* in horses are distinct, it seems wiser to restrict *roan* to the nonprogressive pattern and *grey* to the progressive pattern. This approach is used throughout the book.

Grey

Grey is a pattern of white hairs interspersed with colored body and point hairs. The horse is born colored and then becomes progressively whiter each time it sheds. *Grey* is an exception to the color naming scheme, as the color of the horse is ignored in the name. It is frequently impossible to determine the base color, since the progressive whitening obscures it. On some *greys* the mane and tail become white before the body; on others they remain dark indefinitely. Fig. 93 shows a young *grey* horse that is just beginning the color-changing process. The mane and tail are lighter than the body. The *grey* horse in Fig. 94 also has a light mane and tail but is older and has the typical *dappled* pattern of most *greys*. The horse in Fig. 95 has about the same amount of greying but has a dark mane and tail, and the legs are darker than are often seen. The well-developed dappling and light head help to differentiate animals like this from *roan*. Fig. 96 is an older horse but shows the type of *grey* pattern that never loses color on the legs and mane. Like most aged *greys*, the horse in Fig. 97 has become mostly white, while the aged *grey* mare in Fig. 98 has maintained the dark mane and tail. Even on aged *greys* the skin remains pigmented and shows as a darker color where the hair is thin or absent.

Greys that are born *chestnut* or *sorrel* are called *rose grey* (Figs. 99 and 100) because they never have the truly grey appearance that the *blacks* and *bays* have as they become progressively whiter. Fig. 99 is a *liver chestnut* going grey. This animal will look more *rose grey* later in life. Fig. 100 is the shade most commonly referred to as *rose grey*. Fig. 101 is a *red dun* going grey, for which no special name exists. Since this is a *red dun*, it might look *rose grey* later in life, but the yellower *duns* would look neither *grey* nor *rose grey*.

Special names have been used to describe the various

shades of *grey*, such as "steel" and "porcelain," but they are not of much use, since the horse never remains the same shade for long. A few special names are valid, however. *Iron grey* is used to describe those greys that are not prominently *dappled*, as in Figs. 100, 101, and 102. On many *greys* small flecks of color will grow into the coat, as shown in Figs. 103 and 104, and the result is termed *flea-bitten grey*. An extremely rare occurrence is in Fig. 105, which shows patches of red color growing into the coat of a *grey* horse. These patches are called *blood marks* and can become progressively larger, rarely resulting in a very aged *grey* horse that appears uniformly red.

Greying tends to mask any other patterns of white on the horse as the horse ages. The *grey overo* in Fig. 106 demonstrates the diminished contrast between the grey color and the white body patches. The Argentines call this and other "blue"-and-white combinations "*azulejo*," the name of a South American bluebird. In Fig. 107, a horse born *black patterned leopard* shows how much can be hidden by the greying effect.

Roan

Roan can refer to any mixture of colored and white hairs, but it is used here to describe only one pattern in horses. The white hairs are usually limited to the body and do not occur in the mane, tail, legs, or head, although a few breeds that have been selected specifically for the *roan* pattern (such as the Brabant Belgian) do have white hairs on these areas. *Roans* are very seldom dappled. *Roan* is not progressive, as is greying, so *roan* foals are born *roan* or will shed to *roan* after their foal coats and change little after that. Hair that grows over scars is frequently solid-colored, appearing darker than the rest of the coat (Figs. 116, 117, and 119). Though this pattern is not progressive, it does change throughout the year. The coat is lightest in the spring, as in Fig. 109, looking almost white with dark points. In summer it can become darker, as shown in Fig. 110, the intermediate summer stage. The horse is usually darkest in the winter, sometimes even looking non*roan* (Fig. 111). However, a close inspection reveals that both extremes still contain mixtures of dark and white hairs.

The *roans* can be described by the base color of the horse

followed by the term *roan*, as in *black roan*, *bay roan*, and *red dun roan*. However, some shades of *roan* have special names that are well accepted. Fig. 108 illustrates the roan pattern on a *black* horse, a *blue roan*. Although some people reserve *blue roan* for mixtures of red, black, and white hairs, as would occur with *mahogany bay* or *seal brown*, general usage is that all roan horses with a bluish color are called *blue roans*. This would include *black roans*, *seal brown roans*, and probably some *brown roans*. The *blue roan* pictured here is faintly dappled, which is rare on *roans*. *Purple roan* refers to the roan pattern on *mahogany bay*, which results in a purple color.

Bay roan, as shown in Fig. 109, 110, and 111, is commonly called *red roan*. Figs. 109 and 111 are of Brabant Belgian horses, so the white hairs are normally present in the manes and tails. Fig. 109 was taken in the spring, when *roans* are light in color. Fig. 110 is the intermediate summer stage. Fig. 111 was taken in late fall and appears non*roan* from a distance. *Sorrel roan*, shown in Fig. 112, is commonly called *strawberry roan*. Fig. 113 is another *strawberry roan* with a very dark tail that could be confused with black. *Honey roan* describes the *roan* pattern on light *sorrel alazáns*, especially on *blond sorrels*. *Lilac* or *lavender roan* refers to the roan pattern on dark *chestnut* or *liver chestnut*, as in Fig. 114. No special names have arisen for many combinations, including the roan pattern on a *dark buckskin* (Fig. 115), and so the correct designation is *dark (smutty) buckskin roan*. In Fig. 116 the designation is *smutty olive grullo roan*. Dark hairs have grown in over the branding scar.

Horses of the *roan* pattern sometimes exhibit small dark spots that are unrelated to scarred areas. The number of dark spots usually increases with age. They are not like the spots seen in the *leopard appaloosa*. Since this pattern resembles the arrangement of dark kernels on an ear of Indian corn, the term *corn* was used in the American West to describe the pattern. The *corns* are a further modification of *roans* and can be lumped together with them. The term *corn* simply divides the *roan* classification to reflect this detail of distinctly colored spots. *Blue corn*, Fig. 117, is the counterpart of *blue roan*. *Purple corn*, Fig. 118, is the variation of *purple roan*. The *red corn* in Fig. 119 also illustrates pigmented hairs growing into scarred areas, in this case a brand. Although not heavily spotted, the horse in

Fig. 120 is a *strawberry corn*, and it has a mane and tail darker than those of most *sorrels*.

Rabicano

Rabicano (pronounced rah-bee-CAH-no; also called *squaw tail*) is a pattern that occurs in a large number of breeds, including the Arabian, Quarter Horse, Noriker, Brabant, and North American Spanish, but it is usually missed in most color descriptions of horses. The smallest extent of this pattern occurs when a few white hairs are confined to the flank and the base of the tail, as in Figs. 121 and 122. Extensively marked horses have a pattern of white hairs that seems to extend out from the flank as well as numerous white hairs at the base of the tail, as in Fig. 123. Very extensively marked horses (Fig. 124) are confused with *roan*, but close observation will determine that the pattern is distinct from the usual *roan* pattern.

Frosty

Frosty, or *silvery* (also called *skunk tail*), is a rare pattern that is similar to *rabicano*. The white hairs are at the base of the tail and in the mane, and they can also occur down the back of the horse, over the pelvic bones, and over other bony prominences on the body, such as the hocks. Fig. 125 is a *red bay frosty* in winter coat, showing the light back and tail. Fig. 126 is a *mahogany bay frosty* showing the light roaning over the back and over the pelvic bones. It is probably this pattern, in addition to *roan*, that is responsible for the roan manes and tails on *roan* horses such as the Brabants.

Asymmetric White Spotting Patterns

Asymmetric patterns of white spotting are called "paint" or "pinto" patterns in the United States and consist of irregular patches of white on any base color. The patches usually involve well-defined areas of solid white as opposed to a mixture of white and colored hairs like that which occurs on roans and greys. Several different patterns (*tobiano*, *overo*, *sabino*, *splashed white*) fall into this group, and each is a separate and distinct pattern. Some occur only in a few breeds.

Piebald and *skewbald* are other terms that have been used to describe horses having any of the asymmetrical white patterns. *Piebald* refers to a *black* horse with any of these white spotting patterns, since *piebald* derives from "magpie," a black-and-white bird. *Skewbald* refers to a non*black* horse with any of these patterns. Both terms originated in Britain, where white spotting is rare on horses. Just as the term *dun* glosses over a multitude of colors, the terms *piebald* and *skewbald* also ignore which specific pattern is present. It is more accurate to state the background color of the horse and then the specific pattern that is present.

Tobiano

The *tobiano* (pronounced toe-bee-AH-no) pattern is the most common type of white spotting seen on horses in the United States. In this pattern the white areas usually have a distinct, sharp edge to them. The topline (from ears to tail) is usually crossed at some point by a white patch. All four lower legs are usually white, and the head is usually colored and patterned with the conservative markings common on unspotted horses. The eyes, as a rule, are not blue. *Tobiano* can range from very little white spotting (perhaps limited to a white patch along the topline and four white legs) to a horse largely white (usually with the body white and the head colored). Fig. 127 is a *brown tobiano* with minimal spotting, and Fig. 128 is a *mahogany bay tobiano* with little spotting. Since the horse in Fig. 129 is a *black tobiano* (with an intermediate level of spotting) it could also be called *piebald*. Note that the facial markings of this horse are quite conservative. Fig. 130 is a *chestnut alazán tobiano* with the middle range of marking; it is unusual in that one fetlock and hoof are colored. Figs. 131 and 132 are *bay tobianos* with more extensive marking.

Overo

The *overo* (pronounced oh-VAY-roh) pattern is less common in the United States than the *tobiano*. The white areas usually display more ragged edges than do the *tobiano* spots, and it is rare for white on an *overo* to extend onto the topline except on horses with a great deal of white. The head is usually marked extensively with white, and the eyes are frequently, but not always, blue. The white body patches commonly occur in the middle of the sides and neck, or they may appear on the belly. As a result, the general impression is usually a horizontally arranged white pattern, as opposed to the *tobiano* pattern, which appears more vertically arranged. *Overos* often have one or more colored feet. A *bald face* and a small white spot on one side (these small spots frequently have a butterfly shape) constitute the minimum extent of the *overo* pattern (Fig. 133). A white head, white body, and spots of color peripherally on the feet or the topline constitute the maximum extent of the pattern (Fig. 137). *Overos* are rarely all white, since most of the white foals die a few days after birth because of intestinal malfunctions.

The horse in Fig. 133 is a *red bay overo* with minimal spotting, while Fig. 134 is a *sorrel overo* with a medium grade of spotting. The spots remain in the middle of the neck and body. Fig. 135 is a *chestnut ruano roan (lilac roan) overo*. Although this horse is not intensely roaned, we should remember that the roaning is separate from the *overo* pattern. Fig. 136 is a *red bay overo rabicano*. The *rabicano* did not result in roan edges around the spots, although in some horses it does (Fig. 138). This horse is more extensively marked, and white has crossed the topline to include the mane. The *red bay overo* in Fig. 137 shows all the hallmarks of an *overo*: dark legs, white head, and dark topline. Fig. 138, a *black overo rabicano*, is extensively marked and has white in the mane; the edges of the spots are roan. Even though the horse is extensively marked, the front feet remain pigmented. As this is a black-and-white horse, *piebald* would also be a correct designation.

Sabino

Sabino (pronounced sah-BEE-no) is a fairly rare pattern in the United States. It is usually not identified as a separate pattern but is erroneously lumped with *overo*. It is a distinct pattern and is sometimes called *flecked roan, buttermilk roan, calico paint*, or *roan* (*roan* is the designation usually used in the descriptions of the Clydesdale breed, though it is incorrect). The white patches are quite variable. On some horses they are very distinct and sharp, but on others only a flecking of small spots of white occurs on the background color. Most animals have both flecks and patches. The head usually exhibits extensive white, although somewhat less than that on the *overos*, and commonly the upper lip is pigmented even in very white individuals. The legs are usually white, although many *sabinos* do have a dark leg or two. The patches usually cover the belly, as shown in Fig. 139. The *sabino* pattern is similar to the roan pattern of Shorthorn cattle.

The minimum expression of this pattern usually consists of three or four high white stockings and a *bald* or *apron* face. Horses having this minimal expression of the pattern are usually classified as nonspotted. Fig. 140, a *bay sabino*, demonstrates the usual minimal pattern in the Clydesdale breed and has four high white leg markings, one of which extends onto the stifle in an irregular fashion. A similar pattern is seen in Fig. 141, a *red bay sabino* with flecks of white on the body. The white spot on the one knee is common in sabinos with dark lower legs.

The white displayed in the intermediate expression takes the form of flecks and patches of white on the body in patterns similar to the *overo* pattern but usually much more ragged and irregular, as shown in Fig. 142, a *sorrel alazan sabino*. Flecks, roaning, and white patches are evident, as well as high white leg markings on the rear leg. Fig. 143 is a *black sabino*, a *piebald*, with both patches and roaning. The *bay* and *sorrel sabinos* in Fig. 144 demonstrate the variations in appearance of the more middle range of spotting. Fig. 145 is a *bay sabino* showing the flecking that can occur. The face is exceedingly white also. Flecks and patches of white are also present on the *sorrel sabinos* in Figs. 146 and 147.

The exceedingly white *sabinos* maintain colored ears, a colored chest patch, and maybe a colored patch on the flank and base of the tail. The horse in Fig. 148, a *sorrel sabino*, has an extreme expression of the pattern. Pigment is present in irregular patches over the ears, chest, tail, head, back, and flank. Occasionally solid white foals are born to *sabino* matings, but most of these very white ones will have colored ears.

Splashed white

Splashed white is a pattern that occurs in some European breeds such as the Finnish Draft Horse and the Welsh Pony. The edges of the white areas are crisp and distinct like *tobiano* spots. The four legs are usually white, the head is largely white, and varying extents of the belly are white. The eyes are commonly blue. The *black splashed white* foal in Fig. 149 shows typical markings and the ashen body color so common on *black* foals. In Fig. 150 the eye is blue and the face is extensively white. Fig. 151 is of a *bay splashed white* horse that is extensively marked.

Medicine Hat Paint

Medicine hat paint is a name given to a pattern on horses that are largely white but have color on the ears, or ears and eyes, and colored patches on the chest, flank, and base of the tail. The patch on the ears is considered the "bonnet," and the body patches are like "war shields." The *medicine hat* is not a separate pattern of white spotting, as are the others, but simply a name for a specific arrangement of the colored areas left on a largely white horse. *War bonnet paint* is a similar pattern with color on the ears and very little on the body. Horses with these markings were thought by some tribes of American Indians to be imbued with supernatural powers. *Overos* that are largely white fit this category (Fig. 138), as do some horses on which both the *tobiano* and *overo* patterns are combined (Fig. 152). *Sabinos* that are largely white are consistently this pattern (Fig. 148), as are *sabino-overo* combinations (Fig. 153).

Symmetric Patterns of White

In addition to patterns of individual white hairs (*grey, roan, rabicano, frosty*) and asymmetric patches of white (*tobiano, overo, sabino, slashed white*), horses can exhibit symmetrical patterns of white. In the United States these patterns are usually thought of in connection with the Appaloosa and the Pony of the Americas, but worldwide they occur in a variety of breeds from ponies to draft horses. These patterns sometimes undergo changes during the life of the horse, but generally they remain constant. The Appaloosa breed recognizes six varieties of patterns: *frost, leopard, varnish roan (marble), white blanket, spotted blanket*, and *snowflake*. In the following discussion the patterns are not divided this way, but rather are grouped to represent the range of expression that each of the distinct symmetrical white spotting patterns can exhibit.

Blanket

Blankets of white vary in extent from small ones situated over the croup and hips of the horse to larger ones covering most of the body. Some of these *blankets* have roan edges, blending into the colored areas. Other *blankets* are solid white and sharply defined, and still others blend into the colored areas with flecks of white hairs. It is difficult to cleanly separate the *roan, solid*, or *flecked blankets* into three separate categories, because many *blankets* show characteristics of all three. Fig. 154 is an example of a small *solid blanket* on a *brown* horse, and Figs. 155 and 156 show increasingly larger *solid blankets* (with *dark spots*). Fig. 157 is a light *brown* with a *flecked blanket*. The horse in Fig. 158 is a *sorrel* with a *flecked blanket*, and the intermingling of flecks of color and white shows up at the edge of the *blanket*. A *black* horse with a *roan blanket* is in Fig. 159.

Leopard Spotting

The description *leopard* is used for horses that are all white, or that have extensive *blankets*, with colored spots on the white. These spots can vary in size from a few to several centimeters across. In one of the distinct types of *leopards*, the spots appear to flow out of the flank and over the body of the horse (*patterned leopard*). In the second type, the spots tend to be rounder and do not appear to flow out of the flank (*unpatterned leopard*).

Fig. 160 is an example of a *claybank dun* (notice the forelock) *patterned leopard* showing spots that flow out of the flank. The horse behind the *claybank dun* is an *unpatterned leopard*. The horse in Fig. 161 is a *black patterned leopard*. The thick mane and tail are characteristic of the Noriker Horse, unlike the sparse manes and tails characteristic of the Appaloosa breed. Thus, the pattern and the sparse mane and tail do not necessarily occur together. The *sorrel alazán patterned leopard* in Fig. 162 is interesting because the basic pattern appears to be an extensive *flecked blanket* upon which is superimposed the patterned *leopard spotting*. Each spot has a pale ring around it. This is a rare marking and the name "*varnish marks*" is sometimes used to describe this effect. Fig. 163 is a closeup of a *chestnut patterned leopard*. Occasionally when the color is broken up into the small spots of the leopard pattern, the red- and black-pigmented hairs populate different areas, and the result is some red spots, some black spots, and occasional mixed spots. Fig. 164 is an example of the *unpatterned leopard* on a *chestnut* horse. A *few-spot leopard* (Fig. 165) is almost white but retains a few spots as well as areas of colored skin. Fig. 166 is a *chestnut tostado* with no pattern of white, but a very weak expression of the *leopard spotting* can be seen. *Leopard spots* can occur on horses with no white pattern, such as the *palomino*, but they are not common. Dark spots unrelated to *leopard spotting* are sometimes seen on *chestnut*, *sorrel*, and *palomino* horses.

Varnish Roan

Varnish roan (marble) is a pattern of white that varies from a simple *roan blanket* to white hairs dispersed over the whole horse. It differs from the usual pattern of *roan* in that the head has white hairs and the colored hairs are concentrated over the bony prominences (facial bones, withers, shoulders, knees, stifles, and pelvic bones). These darker areas, as seen in Fig. 167, are called *varnish marks*. This pattern is likely to change with the age of the horse, since many are born solid and develop this pattern later in life. This horse is a *sorrel alazán varnish roan*. Fig. 168 is a more extensively marked *varnish roan*, but the dark areas are still present.

Snowflake

The *snowflake* pattern is one that varies with the age of the horse. It begins as small spots of white, 1 to 3 centimeters (3/8 to

1¼ inches) across that grow into the coat of a colored horse. Although this pattern may not start until the horse is five years old or so, it usually begins much sooner. The white spots can increase in number as the horse ages until the pattern stabilizes at some point. Some horses stabilize in the *snowflake* pattern, as in Fig. 169, but others become very light, appearing white with colored spots a few centimeters across throughout the coat (Fig. 170). This more advanced pattern is then called *speckled*, representing an end point of the *snowflake* pattern and not a completely separate pattern. Although they are usually easy to distinguish, some *speckled* horses can be confused with *flea-bitten greys*.

On some *snowflake* horses the white spots disappear again, resulting in an unmarked horse. Fig. 61 is an *apricot dun* with a few *snowflake* spots. The pattern on this horse never progressed beyond this point. The pattern on the *mahogany bay snowflake* in Fig. 169 regressed after this photo was taken, and now the horse is *brown* with very few spots remaining. Fig. 170 is a *chestnut speckled* in which the *snowflake* pattern was progressive until it stabilized at this point.

Mottled

Mottled refers to small points of white on the muzzle, on the genitalia, and around the eyes of some horses. Although it is commonly seen on horses with the symmetrical patterns of white, mottling may be distinct from them. Fig. 171 shows facial mottling, and the next illustration, Fig. 172, shows mottling over the body in a pattern that does not fit into any other pattern of white.

The symmetrical patterns of white appear in combinations on many horses. This is especially true of the Appaloosa breed, but less so of others such as the Noriker, in which a limited number of the patterns occur. Fig. 173 is a *black varnish roan* with a *solid blanket* and *dark spots*. The *chestnut ruano* in Fig. 174 exhibits a *solid blanket, snowflake, varnish roan*, and *dark spots*. Fig. 175 is a *dark-spotted* and *speckled* combination on *black*. The *varnish roan* and *blanket* patterns are combined on the horse in Fig. 176, resulting in the absence of *varnish marks* over the pelvic bones.

These horses illustrate the difficulty in assigning patterns to some horses. The horse in Fig. 177 is a *blue roan* with a *blanket* and *dark spots*. The dark spots in the blanket are unaffected by the *roan* pattern. In addition to the classic *roan* pattern, this horse also developed the *varnish roan* pattern as he matured. The *mahogany bay varnish roan* in Fig. 178 is unusual because the pattern is incompletely developed and resembles a *roan blanket*. However, the varnish marks over the hips help to differentiate this pattern from the *roan blanket* pattern.

White

Horses that are entirely white are described simply as *white*. A gene responsible for causing *white* results in a solid white horse, usually with brown eyes (Figs. 179 and 180), but *whites* can also result from extremely marked *overos* (most of which die soon after birth), *sabinos*, and *blanket* patterns. All of these have pink skin, as opposed to an aged *grey*, which has dark skin. The *overo*, *sabino*, and *blanket* horses that are white usually have blue eyes. Most white horses have a few spots of pigmented skin. Very light *cremellos* and *perlinos* can appear white, but usually a hint of color is present.

Minor White Marks on the Face and Legs

White marks on the face and legs are independent of body color and patterns and for that reason are described separately. The skin under these marks is unpigmented, adding to the contrast they present. In horses which grey as they age, the markings that were distinct in youth blend with the white hairs of the pattern. Careful observation, however, reveals the pink skin underneath the white hair of the marks and dark skin under the white hair of the greying areas.

Leg Markings

Descriptions of white leg markings should include the type of pattern present, the location of the pattern if it only surrounds or extends up part of a limb, and the limb that is marked.

Fig. 181 illustrates the common white marking patterns on the front and rear limbs. Note that the unpigmented character of white leg markings extends down onto the hoof wherever it occurs, so the hooves could vary from being dark (Fig. 181*A*) to having white stripes (Fig. 181*B*) or to being all white (Fig. 181*C*).

The smallest leg marking worthy of description is a *white mark* (or *white spot*) such as the *white mark* shown on the outside left coronary band (Fig. 181*B*). If this mark is located on the heel of the foot, it is called a *heel* instead of a *white mark*. A *coronet* (Fig. 181 *C*) is a narrow white band surrounding the area immediately above the coronary band. A *white pastern* (Figs. 181*D*, 181*I*) is a marking that includes more than a *coronet* but does not extend to the level of the fetlock. The *fetlock* marking (Figs. 181*E*, 181*J*) includes white covering some part of the bulge of the fetlock joint. When white extends to the middle of the cannon bone, the marking is called a *sock* (Fig. 181*F*), and anything above that is called a *stocking* (Fig. 181*G*, 181*K*). *Stockings* on some horses extend a great distance above the knee or hock (Figs. 181*H*, 181*L*) and are not acceptable markings in certain registries such as that of the American Quarter Horse Association. Care must be taken when describing body colors on these horses, because point color can be concealed by the stocking, as shown in Fig. 15. Dark spots (*ermine marks*) can occur near the coronary band in any of these patterns and are usually accompanied by a dark stripe through the hoof wall at that location.

Face Markings

Facial markings (Fig. 182) come in a variety of shapes and sizes, with patterns frequently being mixed, so a description may include combinations of several patterns. A *star* (Fig. 182*A*) is a white mark on the forehead that varies from a faint white spot to a large irregular area. A *snip* (Fig. 182*B*) is a white marking between, near, or extending into the nostrils. A narrow marking no wider than the flat nasal bones is called a *strip* (Fig. 182*C*) if it occurs in the area between the eyes and nose. Often a *strip* occurs with either or both the *star* and the *snip* (Figs. 182*D*, 182*E*). A *connected star*, *strip*, and *snip* are called a *stripe* (Fig. 182*E*) if they are narrow and a *race* if they wander off to one side. Extensive white covering most of the forehead between the eyes, the entire width of the nasal bones, and the area between the nostrils is a *blaze* (Fig. 182*F*). *Bald-faced* is a term used to describe a wide blaze where the white extends to or around the eyes or nostrils, with most of the upper lip also white (Fig. 182*H*). White covering both lips is called a *white muzzle*, and it often accompanies *blazes* or *bald faces*. When the white spreads out to wrap around and include both upper and lower lips, the horse is said to have an *apron face* (Fig. 182*H*). The term *bonnet* (Fig. 182*I*) is used if only the ears and eyes remain pigmented, and *paper face* is sometimes used to describe a head that is all white. Some facial marks, like that of Fig. 183, are quite unique and do not fit these descriptions.

Other Markings

Skin that has been disturbed from normal growth can show color changes. Heat branding tends to show as a darker mark (Figs. 116, 117, and 119), while freeze branding usually results in white hair (Fig. 184). White hair may also grow in where saddle sores and other tack irritations once were (Fig. 24).

Genetics

Genetics of Colors

The genetic mechanisms that determine the colors of horses have not been completely documented, and as a result any discussion of coat color genetics in horses must not be taken as absolute. A basic knowledge of genetics is assumed and will not be reviewed here. The nomenclature used in this discussion follows Odriozola's and Adalsteinsson's thinking. Although they differ from other authorities in the assignment of the *B* and *E* loci, this is only important to those readers who are already well versed in a knowledge of coat color genetics.

Point Color

The easiest way to begin thinking of coat color genetics is with the genetics of point color. Genetically, color is either black/chocolate brown or red/flaxen, even though all of the color names are determined by the black/nonblack scheme, where brown is lumped with nonblack. Both black and the chocolate brown point shades are caused by the same chemical pigment, eumelanin, with the differences in shade resulting from the characteristics of the microscopic granules of the pigment. Red/yellow pigment (the color of *sorrels*, *palominos*, or bodies on *bays*) is different chemically from the black/chocolate and is called phaeomelanin. The realization that black/brown is a result of one chemical and red/yellow another will make some of the genetic mechanisms simpler to understand.

The difference between black and brown points is determined by the *B* locus, in which black, *B*, is dominant to chocolate, *b*. Horses with a *B* allele (a pair of genes at a particular locus) are capable of forming black eumelanin; horses with *bb* alleles are incapable of forming black and can only form chocolate eumelanin. This is also true in the skin, so *bb* animals have brown or pinkish skin, while *BB* or *Bb* animals have the more common dark skin.

The cause of red (phaeomelanic) points is the recessive *ee* allele; the dominante *E* allele allows eumelanin in the points, making them black or brown depending on the allele at the *B* locus (*B* or *bb*). So a *B_E_* horse has black points, a *bbE_* horse has brown points, and *B_ee* or *bbee* horses have red points. (The dashes here and in the genotypes throughout the rest of this chapter represent genes, either dominant or recessive, that have no effect on the combination.) The phaeomelanic (red) points cover a range from very dark (easily confused with brown) to red to off-white flaxen points.

The most common cause of flaxen points is a modification of the red (phaeomelanin) points if eumelanin is not allowed in the points by *E*. The *F* locus is used for this effect, the recessive *f* being flaxen and the dominant *F* being red. So *B_E_F_* has black points, as does *B_E_ff*. Horses *bbE_F_* have brown points, as do *bbE_ff* horses. Horses that are genetically *B_eeF_* have red points, those that are *B_eeff* have flaxen points, those that are *bbeeF_* have red, and those that are *bbeeff*, flaxen. The actual situation here may be more complex, with another locus in addition to the *F* locus being an additional cause of red points being expressed as flaxen.

A second cause of flaxen points is the silver dapple gene, *Z*, which is dominant. The flaxen locus *F*, above, works only on phaeomelanin to cause flaxen points. The silver dapple locus works only on eumelanin (black/brown) to cause flaxen points. The result is that *B_E_Z* and *bbE_Z_* have flaxen points, while *B_eeZ_* and *bbeeZ_* have red points unaffected by the presence of the *Z* gene (unless the animal is also *ff*). Fig. 185 is a *bay silver dapple*, illustrating the lack of effect on the body color and a somewhat confusing effect on point color. In Fig. 186 the same allele had faded mane and tail color, but not the leg color.

These four loci, *B*, *E*, *F*, and *Z*, then, govern the colors of the points of the horse. The *B* locus determines whether eumelanin will be black or brown. The *E* locus determines whether the points will be eumelanic (black or brown) or phaeomelanic (red). The *F* locus (along with perhaps other loci) determines if phaeomelanic points will be red or flaxen. The *Z* locus determines if eumelanic points will be black/brown or flaxen.

The complexity of the situation is made easier by the fact that most breeds only have the *B* allele at the *B* locus, which makes brown points impossible in those breeds. Some exceptions to this are the North American Spanish Horse, the Quarter Horse, and perhaps the Morgan. Most breeds also lack the silver dapple gene, *Z*; this allele is rare in the United States except for the Shetland breed, but in Europe some of the light and heavy breeds, including the Shetland, Icelandic, and Dutch Warmblood, have it. The lack of the *Z* and *b* alleles in most breeds means that the difference between black and nonblack points (which is how the colors are divided by visual appearance) is governed by the *E* locus. In most breeds, this makes black points dominant to nonblack points, with the *F* locus determining whether these nonblack points will be red or flaxen.

Body Color

Body color is governed by loci that interact with the basic point color genes. The *A* locus affects body color of horses with eumelanic points (black/brown) but has no effect on the body color of phaeomelanic points (red). The dominant allele, *A*, is responsible for the *bay* and *brown* horse colors. In these shades the points are eumelanic (black) and the body is a phaeomelanic red or brown (a flat brown that is distinct from the chocolate of the eumelanic brown responsible for the color of the points on brown-pointed horses). The differences between *bay* and *brown* are probably governed by independent modifiers that have not as yet been characterized. The fact that *bay* and *brown* are caused by the same gene causes some horsemen to lump the two together; however, since differences can be easily seen, it seems wisest to classify them separately in the hopes that future research may uncover the genetic differences between the two. Horses that have the *A* allele with brown points, *A_bb*, have red bodies and brown points. These are classified visually as *sorrel* or *chestnut* depending on the shade of body red. The recessive allele, *aa*, causes the eumelanin to cover the whole horse, resulting in *black* if the horse is *aaB_* or uniform chocolate browns (*chocolate chestnut*) if it is *aabb*. The a^t allele (unaccepted by

3. Color: horses of several colors—*chestnut alazán*, Milkwood Farms Erica; *sorrel ruano*, Cassy; *sandy bay*, Milkwood Farms Encore; *red roan*, Wait and See; *red bay*, Milkwood Farms Marisa; *red roan*, Fire and Ice; *red roan*, Milkwood Farms Dragon
Breed: Brabant and American Belgian
Owner: Henry and Anne Harper
Photographer: D. P. Sponenberg

4. Color: *black* with blacker points
Owner: Virginia Polytechnic Institute and State University
Photographer: D. P. Sponenberg
Genotype: *aaB_CCddE_*

5. Color: *black* with blacker points; small star
   Breed: Noriker
   Owner: Austrian government
   Photographer: D. P. Sponenberg
   Genotype: $aaB\_CCddE\_$

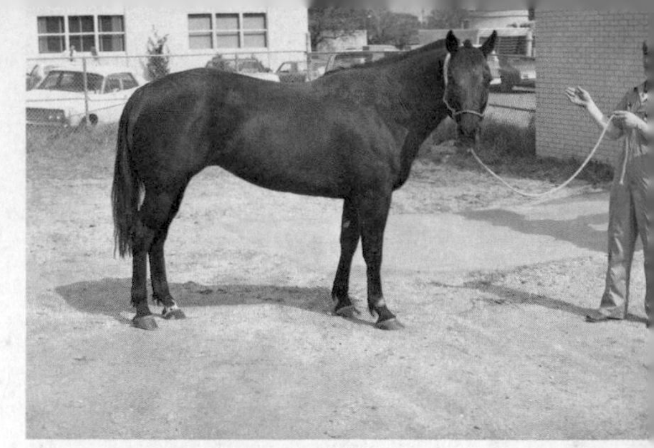

6. Color: faded *black*; half pastern with ermine marks on left rear leg
   Breed: Quarter Horse
   Photographer: D. P. Sponenberg
   Genotype: $aaB\_CCddE\_$

7. Color: *black*, with faded mane and tail; fetlock on left rear, pastern on right rear
   Owner: Jim Price
   Photographer: D. P. Sponenberg
   Genotype: $aaB\_CCddE\_$

8. Color: *jet black*; fetlock on left rear
   Breed: Quarter Horse
   Photographer: B. Beaver
   Genotype: $\_\_B\_CCddE^{d}\_$

9. Color: *smoky black*; pastern on left rear. Foal: *buckskin*; sock on right rear
   Breed: Welsh Pony
   Name: Kingsmead Regalia
   Photographer: J. K. Wiersema
   Genotype: $aaB\_Cc^{cr}ddE\_$

10. Color: *dark brown*; blaze; socks on both rear legs and right front
    Breed: Shire
    Name: Babe
    Owner: Harold Pelky
    Photographer: D. P. Sponenberg
    Genotype: $A\_B\_CCddE\_$

11. Color: *brown*; blaze; stockings on right front and both rear legs
Breed: Clydesdale
Owner: Anheuser Busch Brewing Co.
Photographer: D. P. Sponenberg
Genotype: *A_B_CCddE_*

12. Color: *brown*; fetlock on right rear
Breed: Quarter Horse
Photographer: B. Beaver
Genotype: *A_B_CCddE_*

13. Color: *seal brown*; star and connected strip; pastern on left rear, fetlock on right front
Breed: Spanish Mustang
Name: Ira's Possum
Owner: Tom and Marye Ann Thompson
Photographer: Marye Ann Thompson
Genotype: *aaB_CCddE_P_*

14. Color: *seal brown*
Breed: Spanish Mustang
Owner: Kim Kingsley
Photographer: Kim Kingsley
Genotype: *aaB_CCddE_P_*

15. Color: In front: *seal brown*; bald face; stockings on all four legs.
Behind: *jet black*; blaze; stocking on right rear
Breed: Shire
Name: Flower
Owner: H. Wilson
Photographer: J. K. Wiersema
Genotype: *A_B_CCddE_stysty*; behind, *__B_CCddE*^{d}*_*

16. Color: *red bay*
Breed: Quarter Horse
Photographer: B. Beaver
Genotype: *A_B_CCddE_stysty*

17. Color: *red bay*; star; coronet on left rear, half pastern on right rear
Breed: Arabian
Photographer: D. P. Sponenberg
Genotype: *A_B_CCddE_stysty*

18. Color: *mahogany bay*
Breed: American Indian Horse
Name: Cheyenne Autumn
Owner: Buddy and Leana Rideout
Photographer: D. P. Sponenberg
Genotype: *A_B_CCddE_Sty_*

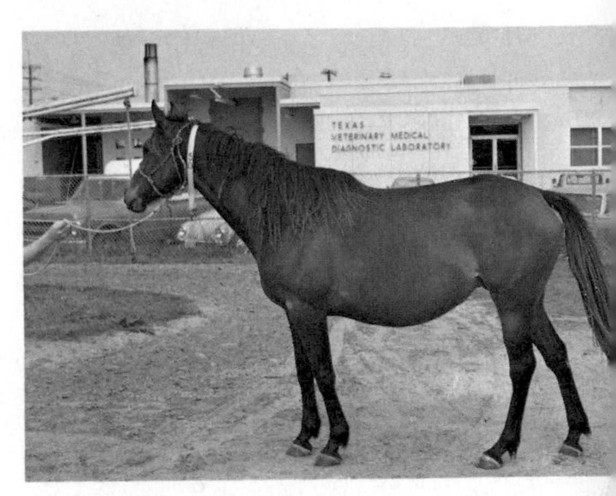

19. Color: *mahogany bay* with faded mane and tail
Breed: Quarter Horse
Photographer: D. P. Sponenberg
Genotype: *A_B_CCddE_Sty_*

20. Color: *blood bay*
Owner: Texas A&M University
Photographer: D. P. Sponenberg
Genotype: *A_B_CCddE_Stysty*

21. Color: *sandy bay*
Breed: American Belgian
Name: Milkwood Farms Encore
Owner: Henry and Anne Harper
Photographer: D. P. Sponenberg
Genotype: *A_B_CCddE_stystyP_*

22. Color: *slate grullo*; race off to right nostril
Breed: Spanish Mustang
Name: Galadriel
Owner: Kim Kingsley
Photographer: Kim Kingsley
Genotype: *aaB_CCD_E_stysty*

23. Color: *slate grullo*; fetlock on right rear, coronet on left rear
Breed: Spanish Mustang
Name: Little Mex
Owner: Lowell and Sharon Scheikofsky
Photographer: D. P. Sponenberg
Genotype: *aaB_CCD_E_stysty*

24. Color: *lobo dun* with white saddle galls
Photographer: D. P. Sponenberg
Genotype: *aaB_CCD_E_Stysty*

25. Color: *olive grullo*
Breed: Spanish Mustang
Name: Ajama de Anza
Owner: Tom and Marye Ann Thompson
Photographer: Marye Ann Thompson
Genotype: *aaB_CCD_E_stysty*

26. Color: *smutty olive grullo*
Breed: Spanish Mustang
Name: Grey Cloud
Owner: Kim Kingsley
Photographer: Marye Ann Thompson
Genotype: $aaB\_Cc^{cr}D\_E\_Sty\_$

27. Color: *silver grullo*; blaze; stockings on both rear legs, half pastern on right front
    Breed: Quarter Horse
    Photographer: D. P. Sponenberg
    Genotype: $aaB\_c^{cr}c^{cr}D\_E\_$

28. Color: *buckskin*
    Breed: Spanish Mustang
    Name: Zi-Chis-Ti-La
    Owner: Lowell and Sharon Scheikofsky
    Photographer: D. P. Sponenberg
    Genotype: $A\_B\_Cc^{cr}ddE\_stysty$

29. Color: *zebra dun*
    Breed: Spanish Mustang
    Name: Xochitl
    Owner: Tom and Marye Ann Thompson
    Photographer: Marye Ann Thompson
    Genotype: $A\_B\_CCD\_E\_stysty$

30. Color: *dusty dun*
Breed: Quarter Horse
Photographer: B. Beaver
Genotype: *A_B_CCD_E_stysty*

31. Color: *dark buckskin*; blaze; stockings on all four legs
Breed: half Arab
Name: Manfred
Owner: H. J. Folmer
Photographer: D. P. Sponenberg
Genotype: $A\_B\_Cc^{cr}ddE\_Sty\_$

32. Color: *silver buckskin*; large star and connected strip, snip; fetlock on left rear, sock on right rear
Breed: Spanish Mustang
Name: Pisgah
Owner: Dan Jones
Photographer: Marye Ann Thompson
Genotype: $A\_B\_Cc^{cr}ddE\_stysty$

33. Color: *silver dun*; fetlock on left rear
Breed: Spanish Mustang
Name: Smoky
Owner: Rev. W. Russow
Photographer: Rev. W. Russow
Genotype: $A\_B\_Cc^{cr}D\_E\_stysty$

34. Color: *chocolate chestnut*; star, snip, and connected strip; heel on left rear
Breed: Morgan
Owner: Virginia Polytechnic Institute and State University
Photographer: D. P. Sponenberg
Genotype: *aabbCCddEE*

35. Color: *liver chestnut tostado*, star
Breed: Spanish Mustang
Name: Chief Cochise
Owner: Dave Tooker
Photographer: Marye Ann Thompson
Genotype: $\_\_\_\_CCddeef\_Sty\_$

36. Color: *liver chestnut tostado*
Breed: Quarter Horse
Photographer: B. Beaver
Genotype: $\_\_\_\_CCddeeF\_Sty\_$

37. Color: *liver chestnut alazán*; star and connected strip; coronet on left rear, fetlock on right rear
Breed: Arabian
Name: El Shanti
Owner: Trish and Dallas Craig
Photographer: D. P. Sponenberg
Genotype: $\_\_\_\_CCddeeF\_Sty\_$

38. Color: *liver chestnut alazán*; blaze; stockings on both fronts and right rear
Breed: Quarter Horse
Photographer: B. Beaver
Genotype: $\_\_\_\_CCddeeF\_Sty\_$

39. Color: *liver chestnut ruano*; wide blaze off to right nostril
    Breed: American Belgian
    Name: Alexander King
    Owner: Gielsky Bros.
    Photographer: D. P. Sponenberg
    Genotype: ____*CCddeeffSty_*

40. Color: *liver chestnut ruano*; bald face
    Breed: Spanish Mustang
    Name: Chief Quanah Parker
    Owner: Gilbert Jones
    Photographer: D. P. Sponenberg
    Genotype: ____*CCddeeffstysty*

41. Color: *chestnut tostado*; sock on left rear, fetlock on right rear
    Breed: Quarter Horse
    Photographer: B. Beaver
    Genotype: *__B_CCddeef_Sty_* or *A_bbCCddE_*

42. Color: *chestnut tostado*; star
Breed: Quarter Horse
Name: Sundown's Tyke
Owner: Larry and Joyce Slusher
Photographer: D. P. Sponenberg
Genotype: *__B_CCddeef_Sty_* or *A_bbCCddE_*

43. Color: *chestnut alazán*; white socks on both rear legs
Breed: Quarter Horse
Photographer: B. Beaver
Genotype: *____CCddeef_Sty_*

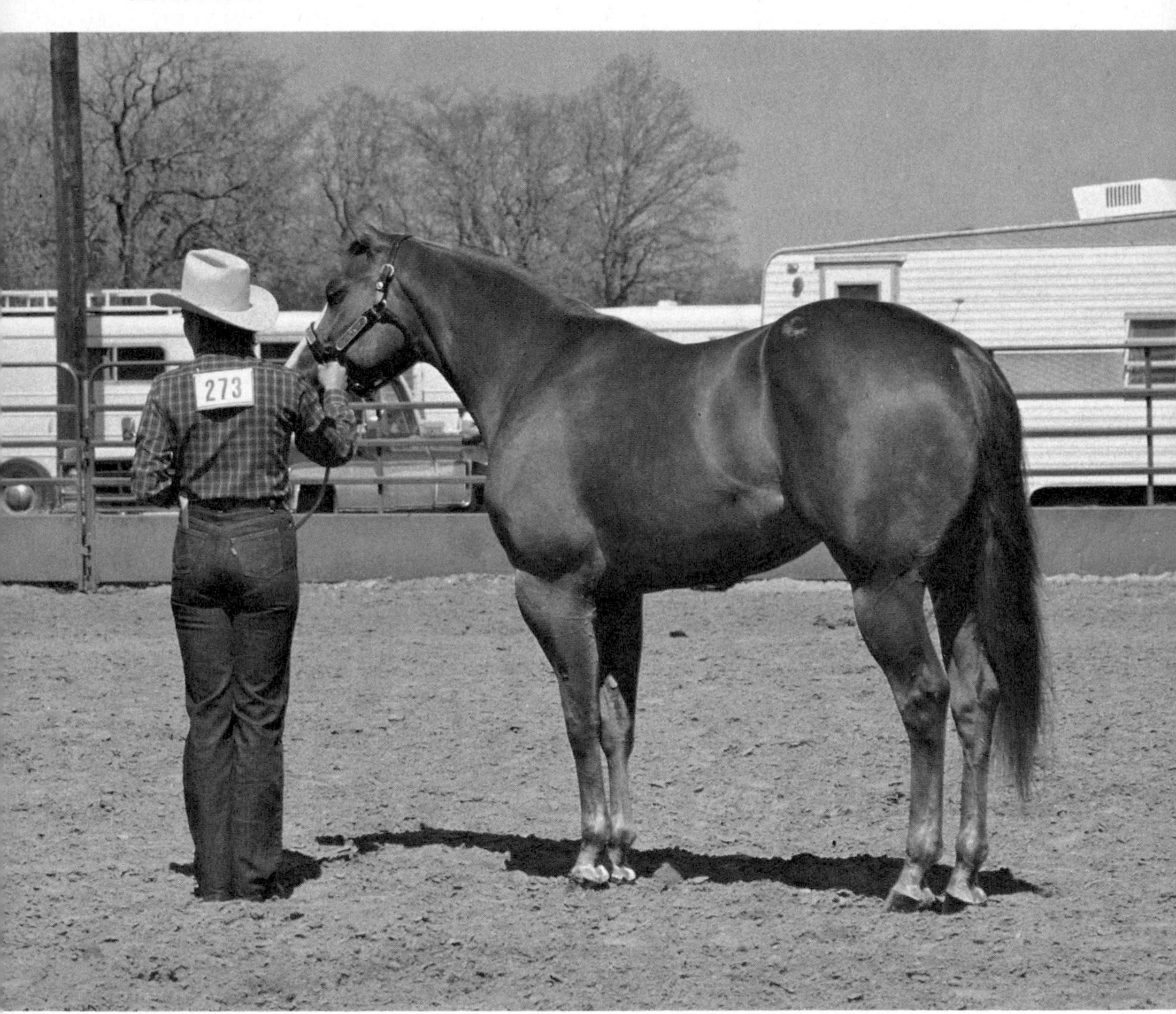

44. Color: *chestnut alazán*; blaze
Breed: Quarter Horse
Photographer: B. Beaver
Genotype: *____CCddeef_Sty_*

45. Color: *chestnut ruano*; blaze; three-quarter stockings on all four legs
Breed: American Belgian
Name: Contractor
Photographer: D. P. Sponenberg
Genotype: ____*CCddeeffSty*_

46. Color: *chestnut ruano*
Breed: American Belgian
Owner: Henry and Anne Harper
Photographer: D. P. Sponenberg
Genotype: ____*CCddeeffSty*_

47. Color: *sorrel tostado*; blaze; sock on left rear
Breed: grade
Name: John
Owner: D. Hislop
Photographer: Debbi Dinnan
Genotype: ____*CCddeeffstysty* or *A_B_CCddE_*

48. Color: *sorrel alazán*; pastern on left rear
Breed: Quarter Horse
Photographer: B. Beaver
Genotype: ____*CCddeeffstysty*

49. Color: *sorrel alazán*; blaze
Breed: Arabian
Photographer: D. P. Sponenberg
Genotype: ____*CCddeeffstysty*

50. Color: *sorrel alazán*; star
Breed: American Belgian
Owner: Henry and Anne Harper
Photographer: D. P. Sponenberg
Genotype: ____*CCddeeffstystyP_*

51. Color: *sorrel alazán*
Breed: Suffolk Punch
Owner: Jane Bukhol
Photographer: D. P. Sponenberg
Genotype: ____*CCddeeffstystypp*

52. Color: *blond sorrel ruano*; star and connected partial strip
Breed: American Belgian
Name: Country Roads Lenore
Owner: Henry and Anne Harper
Photographer: D. P. Sponenberg
Genotype: ____*CCddeeffstystyP*_

53. Color: *lilac dun*; pastern on left rear, half pastern on right rear
Breed: grade pony
Owner: Cornell University
Photographer: D. P. Sponenberg
Genotype: $aabbCc^{cr}ddE\_$

54. Color: *lilac dun*; hazel eyes
Breed: grade
Owner: Cornell University
Photographer: D. P. Sponenberg
Genotype: $aabbCc^{cr}ddE\_$

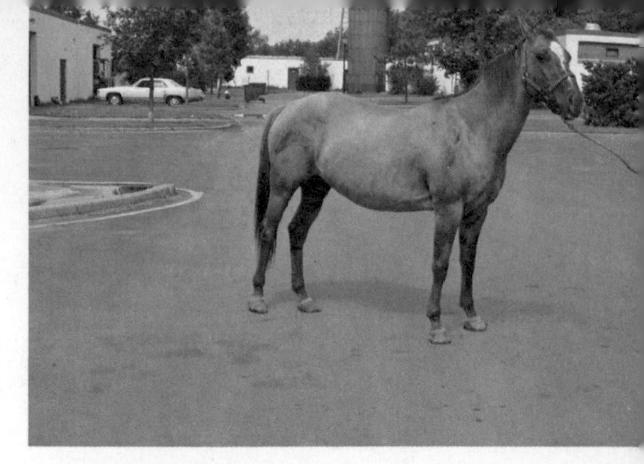

55. Color: *muddy dun*; star and connected strip
Breed: grade
Photographer: D. P. Sponenberg
Genotype: aabbCCD_E_ or ____CCD_eeF_

56. Color: *muddy dun*; star. Foal: *linebacked orange dun*; blaze; fetlock on right rear
Breed: Southwest Spanish Mustang
Name: Red Lightning
Owner: Gilbert Jones
Photographer: Gilbert Jones
Genotype: aabbCCD_E_ or ____CCD_eeF_; foal, ____CCD_eeF_

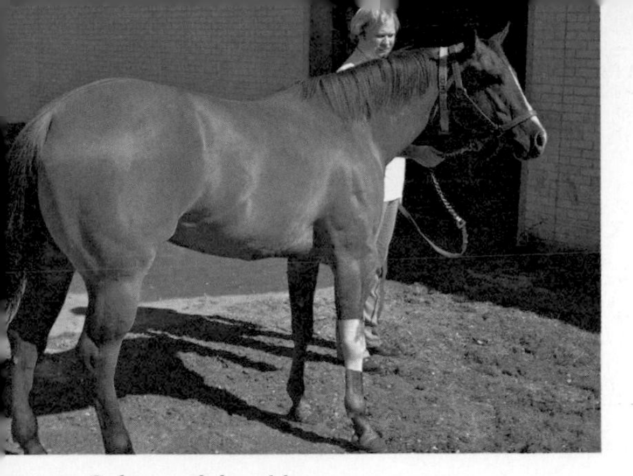

57. Color: *red dun*; blaze
Breed: Quarter Horse
Photographer: D. P. Sponenberg
Genotype: ____CCD_eeF_

58. Color: *red dun*; star and connected strip, separate snip; fetlock on right front
Breed: Quarter Horse
Photographer: D. P. Sponenberg
Genotype: ____CCD_eef_

59. Color: *linebacked orange dun*; three-quarter stocking on left rear
Breed: Quarter Horse
Photographer: B. Beaver
Genotype: ____CCD_eeF_

60. Color: *linebacked apricot dun* in summer coat; star and connected strip; sock on right rear, three-quarter stocking on left rear
Breed: American Indian Horse
Name: Penny's Sundance
Owner: Buddy and Leana Rideout
Photographer: D. P. Sponenberg
Genotype: *____CCD_eef_*

61. Color: *linebacked apricot dun* in winter coat; star and connected strip; sock on right rear, three-quarter stocking on left rear
Breed: American Indian Horse
Name: Penny's Sundance
Owner: Buddy and Leana Rideout
Photographer: D. P. Sponenberg
Genotype: *____CCD_eef_*

62. Color: Left: *linebacked yellow dun*; fetlock on right rear. Right: *zebra dun*
Breed: Spanish Mustang
Name: Sunflower
Owner: Dave and Joyce Smith
Photographer: Joyce Smith
Genotype: left: *A_bbC_D_E_*; right, *A_B_CCD_E_styssty*

63. Color: *claybank dun*; socks on both rear legs
Breed: Quarter Horse
Photographer: B. Beaver
Genotype: ____Cc^{cr}(?)ddeeF_stysty

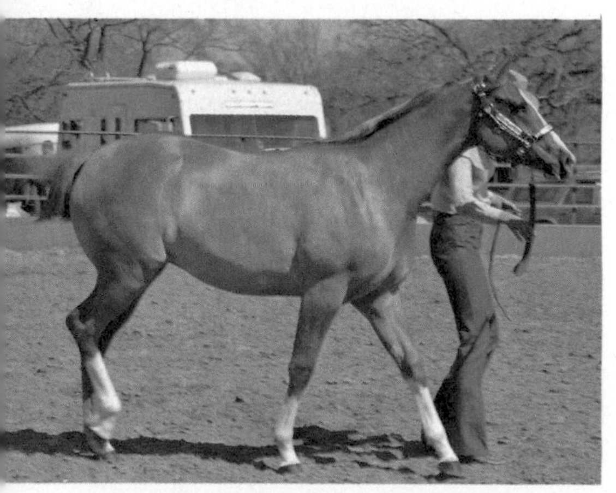

64. Color: *linebacked claybank dun*; blaze; three-quarter stocking on left front leg; sock on right front leg and both rear legs
Breed: Quarter Horse
Photographer: B. Beaver
Genotype: ____Cc^{cr}D_eeF_stysty

65. Color: *linebacked claybank dun* dam and foal
Breed: Quarter Horse
Photographer: B. Beaver
Genotype: ____Cc^{cr}D_eeF_stysty

66. Color: *palomino*; bald-faced
Breed: Quarter Horse
Name: Photostat Mac
Owner: Darrell and Gay Lovell
Photographer: B. Beaver
Genotype: ____$Cc^{cr}ddeestysty$

67. Color: *dappled palomino*; blaze
Breed: Quarter Horse
Name: Jack's Gold Machine
Owner: B. Beaver
Photographer: B. Beaver
Genotype: ____$Cc^{cr}ddeestysty$

68. Color: pink-skinned *palomino*; amber eyes; star
Breed: American Saddlebred
Name: So Proudly We Hail
Owner: Richard Elliott
Photographer: Rick Bate
Genotype: $\_\_bbCc^{cr}ddeestysty$

69. Color: *smutty palomino*; fetlocks on both rear legs
    Breed: Quarter Horse
    Photographer: B. Beaver
    Genotype: $\_\_B\_Cc^{cr}ddeeSty\_$

70. Color: *smutty palomino*
    Breed: Quarter Horse
    Photographer: B. Beaver
    Genotype: $\_\_B\_Cc^{cr}ddeeSty\_$

71. Color: *isabella*; blaze; stockings on both rear legs and left front
    Breed: Quarter Horse
    Name: Imperial Sun
    Owner: B. Beaver
    Photographer: B. Beaver
    Genotype: $\_\_B\_Cc^{cr}ddeestysty$

72. Color: *silver dapple*
    Breed: Shetland Pony
    Name: Jim
    Owner: Nicholas Sharpe
    Photographer: B. Beaver
    Genotype: $aaB\_CCddE\_Z\_$

73. Color: *silver dapple*; star; pastern on right rear, fetlock on left rear
    Breed: Dutch Warmblood
    Photographer: D. P. Sponenberg
    Genotype: $aaB\_CCddE\_Z\_$

74. Color: *cremello*; blaze; stockings on both rear legs
    Name: Wizard
    Owner: Shannon Roberts
    Photographer: Doug Gregg
    Genotype: $\_\_\_\_c^{cr}c^{cr}ddee$

75. Color: *perlino*; blue eyes
Breed: Connemara Pony
Name: Arenbosche Rosalinde
Photographer: J. K. Wiersema
Genotype: $A\_B\_c^{cr}c^{cr}ddE\_$

76. Color: *dappled mahogany bay*; heel on outside left rear
Breed: Quarter Horse
Photographer: B. Beaver
Genotype: *A_B_CCddE_Sty_*

77. Color: *dappled black*
Breed: Shire
Photographer: Sue Quick
Genotype: *__B_CCdd*E^{d}*_*

79. Color: *zebra dun* (black dorsal stripe)
Breed: Icelandic
Photographer: J. K. Wiersema
Genotype: *A_B_CCD_E_StySty*

78. Color: *mahogany bay pangaré*; star
Breed: Brabant Belgian
Name: Milkwood Farms Marisa
Owner: Henry and Anne Harper
Photographer: D. P. Sponenberg
Genotype: *A_B_CCddE_Sty_P_*

80. Color: *zebra dun* foal (red dorsal stripe)
Breed: Quarter Horse
Owner: Texas A&M University
Photographer: B. Beaver
Genotype: *A_B_CCD_E_StySty*

81. Color: *dusty dun*
Breed: Quarter Horse
Photographer: B. Beaver
Genotype: *A_B_CCD_E_StySty*

82. Color: *mahogany bay* with wither and neck stripes; star; half pastern on left front and right rear. Dam: *dappled bay grey*
Breed: Welsh Pony
Owner: W. Strenstra
Photographer: J. K. Wiersema
Genotype: *A_B_CCddE_StySty*

83. Color: *grullo* with zebra stripes
Breed: Quarter Horse
Photographer: D. P. Sponenberg
Genotype: *aaB_CCD_E_StySty*

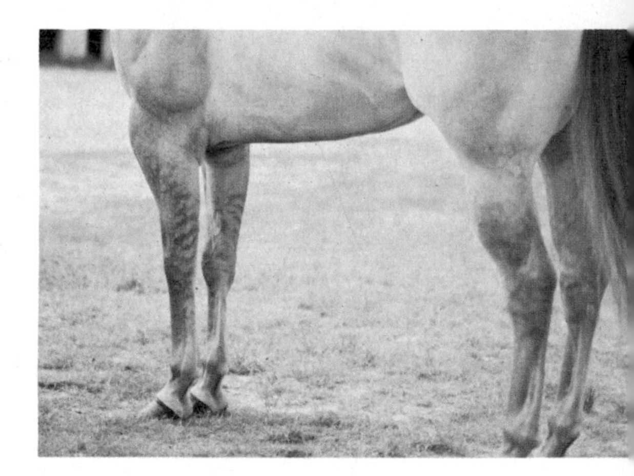

84. Color: *claybank dun* with zebra stripes
Breed: Quarter Horse
Photographer: B. Beaver
Genotype: $\_\_\_\_Cc^{cr}D\_E\_ff$

85. Color: *silver dapple* foal; star
Breed: Dutch Warmblood
Photographer: J. K. Wiersema
Genotype: *aaB_CCddE_Z_*

86. Color: *linebacked bay* foal; star
Breed: Spanish Mustang
Owner: Tom and Marye Ann Thompson
Photographer: Marye Ann Thompson
Genotype: *A_B_CCddE_*

87. Color: *grullo* foal; blaze; pastern on left front, fetlock on right rear; three-quarter stocking on left rear. Dam: *black*
Breed: Spanish Mustang
Owner: Phyllis Falconer
Photographer: Marye Ann Thompson
Genotype: *aaB_CCD_E_*; dam, *aaB_CCddE_*

88. Color: *grullo* foal; star
Breed: American Indian Horse
Owner: Buddy and Leana Rideout
Photographer: D. P. Sponenberg
Genotype: *aaB_CCD_E_StySty*

89. Color: dark eye on *white* horse
Photographer: Alda Buresh

90. Color: amber eye on *palomino*
Photographer: B. Beaver

91. Color: blue eye on *chestnut* (pigmented skin)
Photographer: D. P. Sponenberg

92. Color: blue eye on *silver grullo*
Photographer: D. P. Sponenberg

93. Color: *grey* with light mane and tail, born *black*
Breed: Quarter Horse
Photographer: B. Beaver
Genotype: *aaB_CCddE_G_*

94. Color: *dappled grey* with light mane and tail, born *bay*
Breed: half Arabian
Name: Frosted Flakes
Owner: Angie Bugg
Photographer: D. P. Sponenberg
Genotype: *A_B_CCddE_F_G_*

95. Color: *dappled grey* with dark points, born *black*
Breed: Quarter Horse
Photographer: B. Beaver
Genotype: *G_*

96. Color: *grey* with dark mane and legs, born *black*; fetlocks on both rear legs and left front, three-quarter stocking on right front
Breed: American Indian Horse
Name: Wolf
Owner: Buddy and Leana Rideout
Photographer: D. P. Sponenberg
Genotype: *aaB_CCddE_G_*

98. Color: light *grey*, born ?; dark points
Breed: Quarter Horse
Photographer: B. Beaver
Genotype: *G_*

100. Color: *rose grey* (also *iron grey*), born *chestnut alazán*
Breed: Arabian
Name: Tarah
Owner: Nanci Falley
Photographer: Nanci Falley
Genotype: *____CCddeeG_*

97. Color: light *grey*, born ?; light points
Breed: Quarter Horse
Photographer: B. Beaver
Genotype: *G_*

99. Color: *dappled rose grey*, born *liver chestnut tostado*; socks on both rear legs, three-quarter stocking with ermine mark on left front
Breed: Quarter Horse
Photographer: B. Beaver
Genotype: *____CCddeeSty_G_*

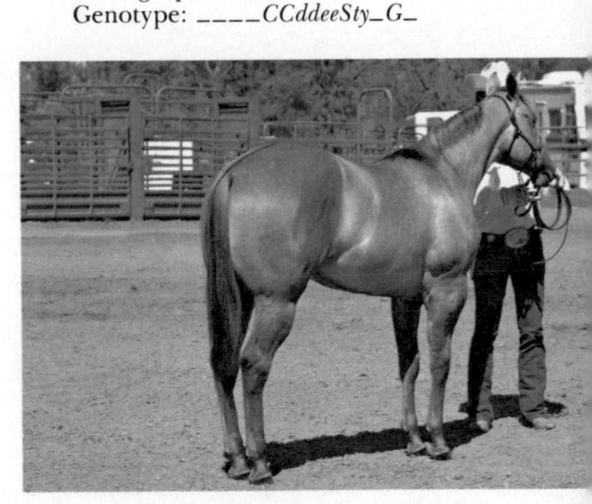

101. Color: *grey*, born *red dun*
Breed: Quarter Horse
Photographer: B. Beaver
Genotype: *____CCD_eeG_*

102. Color: *grey*, born *bay*; pastern on right front
Owner: Sue Tucker
Photographer: D. P. Sponenberg
Genotype: *A_B_CCddE_StyStyG_*

103. Color: *flea-bitten grey* with dark points
Breed: Arabian
Photographer: D. P. Sponenberg
Genotype: *G_*

104. Color: *flea-bitten grey* with dark points
Breed: Arabian
Photographer: D. P. Sponenberg
Genotype: *G_*

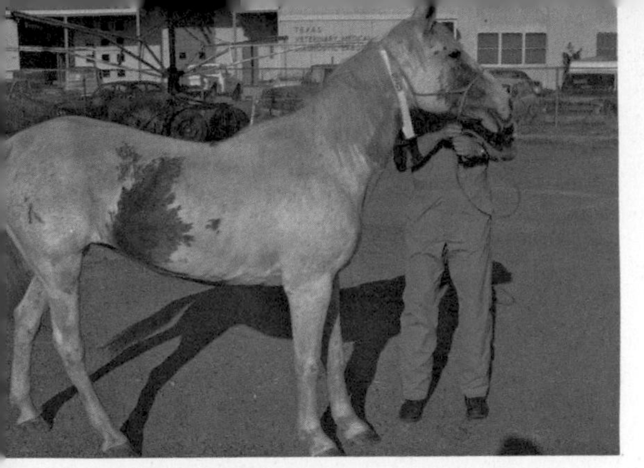

105. Color: *blood-marked grey*
Photographer: D. P. Sponenberg
Genotype: *G_*

106. Color: *grey overo*, born *black overo*; blaze with white muzzle
Breed: American Indian Horse
Name: Choctaw Ghost Dancer
Owner: Nanci Falley
Photographer: Nanci Falley
Genotype: *aaB_CCddE_G_OO*

107. Color: *grey*, born *black patterned leopard*
Breed: Appaloosa
Owner: H. J. Folmer
Photographer: D. P. Sponenberg
Genotype: *G_Lp_*

108. Color: *blue roan (black roan)*
Breed: Quarter Horse
Photographer: D. P. Sponenberg
Genotype: *aaB_CCddEE_Rr*

109. Color: In front: *red roan (bay roan)* in spring coat. Behind: *sorrel ruano*; blaze
Breed: Brabant Belgian
Name: Milkwood Farms Dragon
Owner: Henry and Anne Harper
Photographer: D. P. Sponenberg
Genotype: *A_B_CCddE_Rr*; behind, *____CCddeeffstysty*

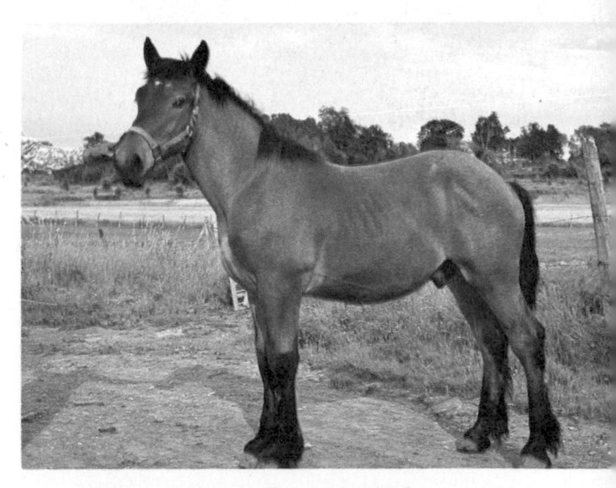

110. Color: *red roan (bay roan)* in late summer coat; star
Breed: American Belgian
Name: Milkwood Farms Mitzi
Owner: Henry and Anne Harper
Photographer: D. P. Sponenberg
Genotype: *A_B_CCddE_Rr*

111. Color: *red roan (bay roan)* in winter coat
Breed: Brabant Belgian
Name: Milkwood Farms Mitzi
Owner: Henry and Anne Harper
Photographer: D. P. Sponenberg
Genotype: *A_B_CCddE_R_*

112. Color: *strawberry roan (sorrel alazán roan)*; star and connected strip; three-quarter stockings on right front and both rear legs, pastern on left front
Breed: Spanish Mustang
Name: Khalid
Owner: Tom and Marye Ann Thompson
Photographer: Marye Ann Thompson
Genotype: *____CCddeeF_Rr*

113. Color: *strawberry roan (sorrel tostado roan)*; stripe; half-pasterns on left front and left rear, sock on right rear
Breed: Quarter Horse
Photographer: B. Beaver
Genotype: ____CCddeef_StyStyRr

114. Color: *lilac roan (chestnut alazán roan)*; blaze; stockings on both rear legs, coronet on right front
Breed: American Belgian
Owner: Ira Long
Photographer: Doug Gregg
Genotype: ____CCddeef_Sty_Rr

115. Color: *dark buckskin roan*
Breed: Welsh Pony
Name: Hemeljynks Maiflower
Photographer: J. K. Wiersema
Genotype: A_B_Cc^{cr}ddE_Sty_Rr

116. Color: *smutty olive grullo roan* with brand
Breed: Spanish Mustang
Name: Azulita
Owner: Tom and Marye Ann Thompson
Photographer: Marye Ann Thompson
Genotype: aaB_CCD_E_Sty_R_

117. Color: *blue corn (black corn)* with brand
Breed: Spanish Mustang
Name: Four Lane
Owner: Wild Horse Research Farm
Photographer: Marye Ann Thompson
Genotype: aaB_CCddE_Rr

118. Color: *purple corn (mahogany bay corn)*
Breed: Welsh Pony
Name: Roman Tinkerbell
Photographer: J. K. Wiersema
Genotype: A_B_CCddE_Sty_Rr

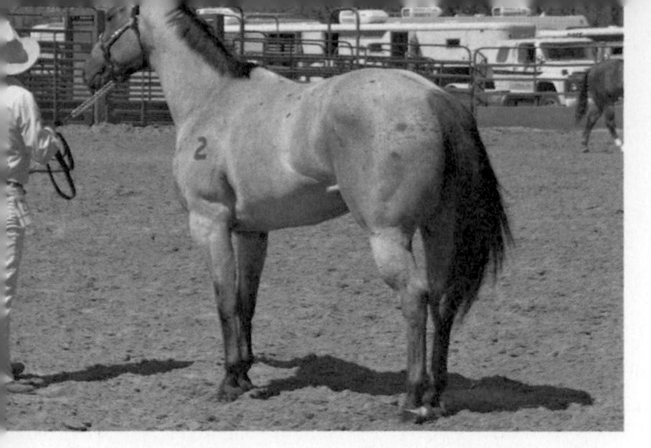

119. Color: *red corn (red bay corn)* with brand; heel coronet on left rear
Breed: Quarter Horse
Photographer: B. Beaver
Genotype: *A_B_CCddE_StyStyRr*

120. Color: *strawberry corn (sorrel tostado corn)*
Breed: Spanish Mustang
Name: Sonora
Owner: Tom and Marye Ann Thompson
Photographer: Marye Ann Thompson
Genotype: *____CCddeef_StyStyRr*

121. Color: *buckskin rabicano*; star; fetlock on right rear, sock on left rear
Breed: Quarter Horse
Name: Imperial Skip
Owner: E. F. ("Bud") Alderson
Photographer: B. Beaver
Genotype: $A\_B\_Cc^{cr}ddE\_StyStyRb\_$

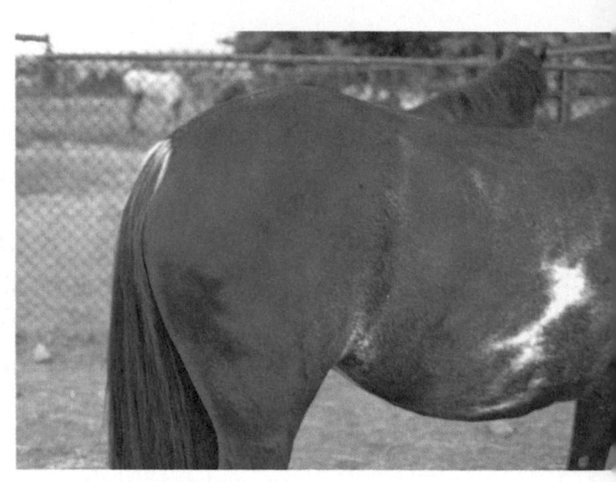

122. Color: *black overo rabicano*
Owner: Cornell University
Photographer: D. P. Sponenberg
Genotype: *aaB_CCddE_Rb_*

123. Color: *chestnut alazán rabicano*; blaze; sock on left front, three-quarter stockings on both rear legs
Breed: Quarter Horse
Photographer: B. Beaver
Genotype: *____CCddeef_Sty_Rb_*

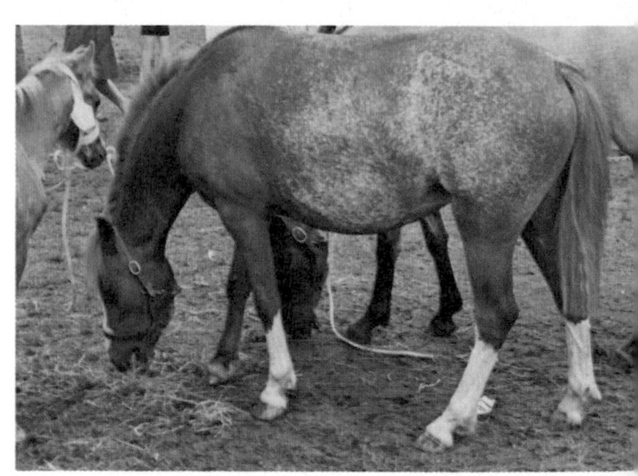

124. Color: In front: *sorrel alazán rabicano*; blaze; three-quarter stockings on both rear legs and left front. Behind: *blue roan (black roan)*
Breed: Welsh Pony
Name: Bellefleur Rosemarye
Photographer: J. K. Wiersema
Genotype: *____CCddeeffStyStyRb_*; behind, *aaB_CCddE_Rr*

125. Color: *red bay frosty*
Breed: Spanish Mustang
Name: Blarney
Owner: Buddy and Leana Rideout
Photographer: D. P. Sponenberg
Genotype: *A_B_CCddE_StySty* and *frosty*

126. Color: *mahogany bay frosty*
Owner: Virginia Polytechnic Institute and State University
Photographer: D. P. Sponenberg
Genotype: *A_B_CCddE_Sty_* and *frosty*

127. Color: *brown tobiano*; high fetlocks on right front and left rear, stockings on left front and right rear
Owner: Virginia Polytechnic Institute and State University
Photographer: Doug Gregg
Genotype: *A_B_CCddE_T_*

128. Color: *mahogany bay tobiano*; blaze; extensive markings on all four legs, ermine marks on outside of left front and inside right rear
Photographer: D. P. Sponenberg
Genotype: *A_B_CCddE_Sty_T_*

129. Color: *black tobiano (piebald)*; stripe; extensive markings on all four legs
Breed: American Indian Horse
Name: Kiowa Sala Li
Owner: Buddy and Leana Rideout
Photographer: Nanci Falley
Genotype: *aaB_CCddE_T_*

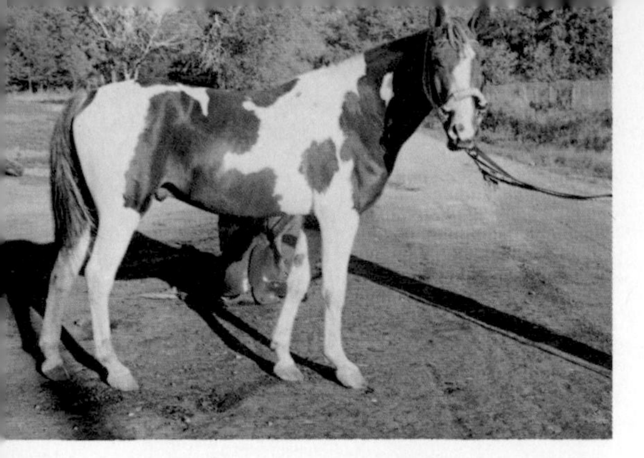

130. Color: *chestnut alazán tobiano*; blaze; white legs
Breed: Southwest Spanish Mustang
Name: Chief Quanah Parker II
Owner: Gilbert Jones
Photographer: Nancy Falley
Genotype: *____CCddeef_Sty_T_*

131. Color: *red bay tobiano*; blaze; sock on left rear, stocking on right rear, extensive white on both front legs, multiple ermine marks on all four feet
Photographer: B. Beaver
Genotype: *A_B_CCddE_StyStyT_*

132. Color: *red bay tobiano*; wide blaze; white legs
Photographer: D. P. Sponenberg
Genotype: *A_B_CCddE_StyStyT_*

133. Color: *red bay overo*; blaze; stocking on left front, high fetlock on left rear
Breed: Spanish Mustang
Name: Yolkai Lin
Owner: Sharon Scheikofsky
Photographer: Marye Ann Thompson
Genotype: *A_B_CCddE_OO*

134. Color: *sorrel alazán overo*; bald face
Breed: Spanish Mustang
Name: Apache Maid
Owner: Jim and Sharon Babbit
Photographer: Marye Ann Thompson
Genotype: *____CCddeeF_stystyOO*

135. Color: *lilac roan overo (chestnut ruano roan overo)*; blaze
Breed: Southwest Spanish Mustang
Name: Look See
Owner: Pauline Johnston
Photographer: Nanci Falley
Genotype: *____CCddeeffSty_R_OO*

136. Color: *red bay overo rabicano*
Breed: Spanish Mustang
Name: Oroneeka
Owner: Karla Davis
Photographer: Marye Ann Thompson
Genotype: *A_B_CCddE_StyStyRb_OO*

137. Color: *red bay overo*; bonnet face
Breed: Southwest Spanish Mustang
Name: Chief of the Choctaws
Owner: Gilbert Jones
Genotype: *A_B_CCddE_StyStyOO*

138. Color: *black overo rabicano (piebald) (medicine hat paint)*; bonnet face; extensive stockings on both rear legs
Breed: Spanish Mustang
Name: Chief Joseph
Owner: Jim and Sharon Babbit
Photographer: D. P. Sponenberg
Genotype: *aaB_CCddE_Rb_OO*

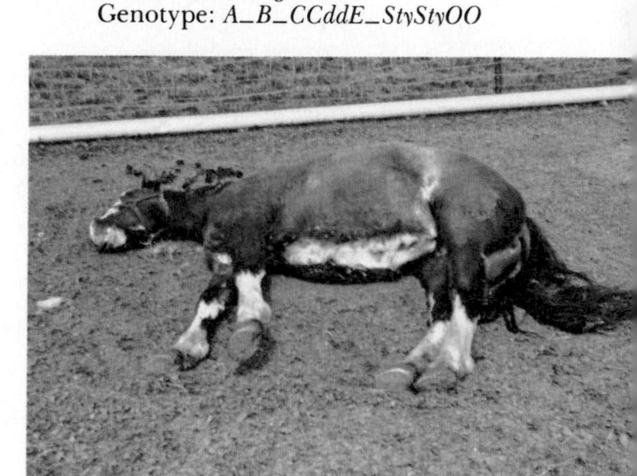

139. Color: *bay sabino*; blaze extending to lower lip; stockings on all four legs
Breed: Spanish Mustang
Owner: Harold Smith
Photographer: Marye Ann Thompson
Genotype: *A_B_CCddE_Sb_*

140. Color: *bay sabino*; blaze; stockings on all four legs
Breed: Clydesdale
Photographer: Sue Quick
Genotype: *A_B_CCddE_StyStySbsb*

141. Color: In front: *red bay sabino*; blaze; stockings on both rear legs, fetlock on right front, white on left knee. Behind: *buckskin*; star
Breed: Spanish Mustang
Name: Shoshoni
Owner: Wild Horse Research Farm
Photographer: Marye Ann Thompson
Genotype: *A_B_CCddE_Sb_*

142. Color: *sorrel alazán sabino*; blaze; sock on left front, extensive stockings on both rear legs
Breed: Spanish Mustang
Name: Booker T. Mustang
Owner: Kim Kingsley
Photographer: Kim Kingsley
Genotype: *____CCddeef_StyStySbsb*

143. Color: *black sabino (piebald)*; wide blaze; extensive stockings on all four legs
Breed: Clydesdale
Photographer: D. P. Sponenberg
Genotype: *__B_CCdd*E^d*_Sbsb*

144. Color: *sorrel alazán sabinos, red bay sabinos*
Breed: Dutch Warmblood
Photographer: J. K. Wiersema
Genotype: sorrel alazán sabino, *____CCddeef_stystySbsb*; red bay sabino, *A_B_CCddE_stystySbsb*

145. Color: *red bay sabino*; wide blaze
Breed: Clydesdale
Name: North River Betty
Owner: David and Mary Flinn
Photographer: D. P. Sponenberg
Genotype: *A_B_CCddE_StyStySbsb*

146. Color: *sorrel alazán sabino*; wide blaze; white legs. Dam: *black*
Breed: Tennessee Walking Horse
Owner: Raymond Hale
Photographer: Doug Gregg
Genotype: *____CCddeef_stystySbsb*; dam, *aaB_CCddE_*

147. Color: *sorrel sabino (medicine hat paint)*; apron face
Breed: Spanish Mustang
Name: Dakota Morning Star
Owner: Jim and Sharon Babbit
Photographer: Marye Ann Thompson
Genotype: *____CCddeeSb_*

148. Color: *sorrel alazán sabino (medicine hat paint)*; paper face; white legs on right rear and both front
Photographer: D. P. Sponenberg
Genotype: *____CCddeef_stystySbSb*

149. Color: *black splashed white* in foal coat; apron face; extensive stockings on all four legs. Dam: *strawberry roan (sorrel alazán roan)*; blaze; sock on right front, three-quarter stockings on left front and both rear
Breed: Welsh Pony
Photographer: J. K. Wiersema
Genotype: *aaB_CCddE_splspl*; dam, *____CCddeeffstystyRr*

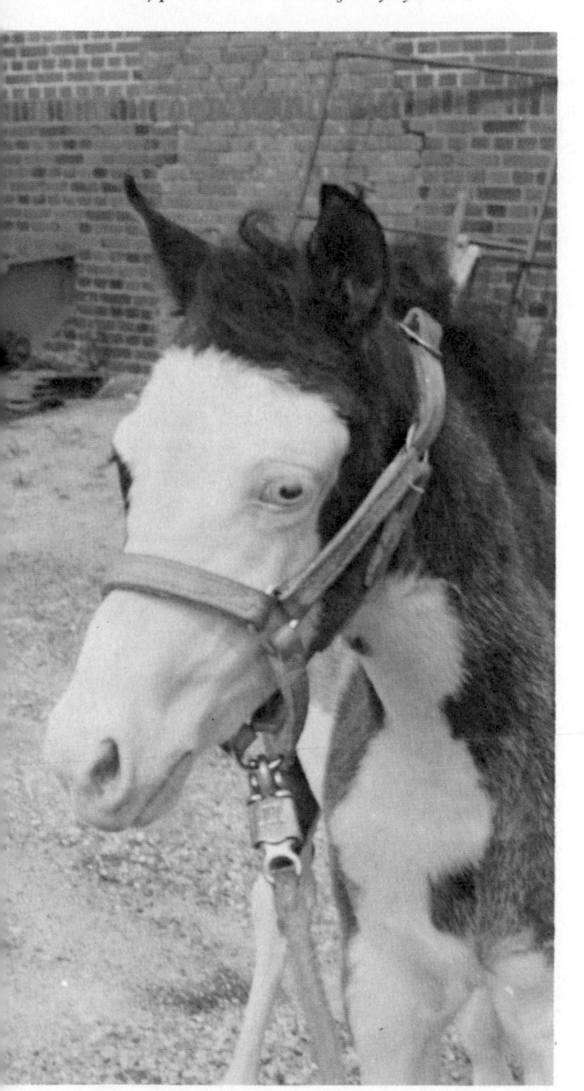

150. Color: *black splashed white*; apron face, blue eye
Breed: Welsh Pony
Photographer: J. K. Wiersema
Genotype: *aaB_CCddE_splspl*

152. Color: *claybank dun overo-tobiano (medicine hat paint)*; bonnet face; white on all legs
Breed: American Indian Horse
Name: Kiowa Morning Song
Owner: Nanci Falley
Photographer: Nanci Falley
Genotype: ____Cc^{cr}*D_eef_stystyT_OO*

151. Color: *red bay splashed white*; apron face, blue eye; white on all legs
Photographer: J. K. Wiersema
Genotype: *A_B_CCddE_StyStysplspl*

153. Color: *overos* and *overo-sabinos (medicine hat paint)*
Breed: Spanish Mustang
Owner: Ferdinand Brislawn
Photographer: Ferdinand Brislawn
Genotype: various colors plus *Sb_OO*

154. Color: *brown* with *solid blanket* and *dark spots*
Breed: Appaloosa
Photographer: B. Beaver
Genotype: *A_B_CCddE_* plus *blanket* plus *dark spots*

155. Color: *bay* with *solid blanket* and *dark spots*; star. Dam: *buckskin*; blaze; stockings on both rear legs, pastern on right front
Breed: Spanish Mustang
Owner: Karla Davis
Photographer: Marye Ann Thompson
Genotype: *A_B_CCddE_*; dam, *A_B_Cc*^{cr}*ddE_stysty*

156. Color: *black* with *solid blanket* and *dark spots*; pastern on right rear
Breed: Appaloosa
Photographer: B. Beaver
Genotype: *aaB_CCddE_* plus *blanket* plus *dark spots*

157. Color: light *brown* with *flecked blanket*
Breed: Appaloosa
Photographer: B. Beaver
Genotype: *A_B_CCddE_stysty* plus *blanket*

158. Color: *sorrel alazán* with *flecked blanket* and *dark spots*; bald face; three-quarter stockings on right rear and right front, stocking on left front
Breed: Appaloosa
Photographer: B. Beaver
Genotype: *____CCddeeffstysty* plus *blanket* plus *dark spots*

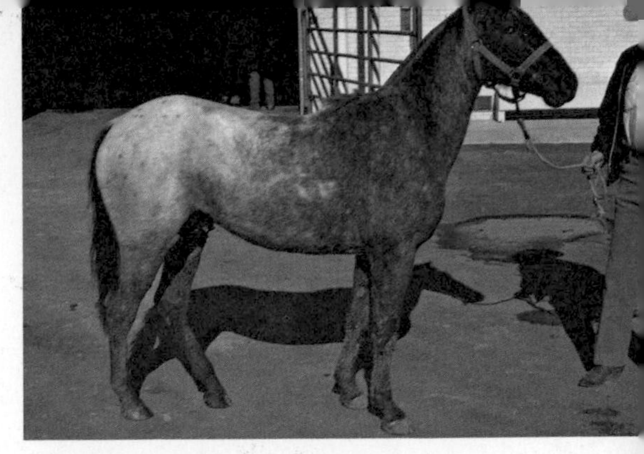

159. Color: *black* with *roan blanket*
Breed: Appaloosa
Photographer: D. P. Sponenberg
Genotype: *aaB_CCddE_* plus *blanket*

160. Color: In front: *patterned leopard claybank dun*. Behind: *unpatterned leopard bay*
Name: Maaike van Koolveld
Owner: O. Lise
Photographer: J. K. Wiersema
Genotype: *____Cc*^{cr}*ddeeffstystyLp_*; behind, *A_B_CCddE_Lp_*

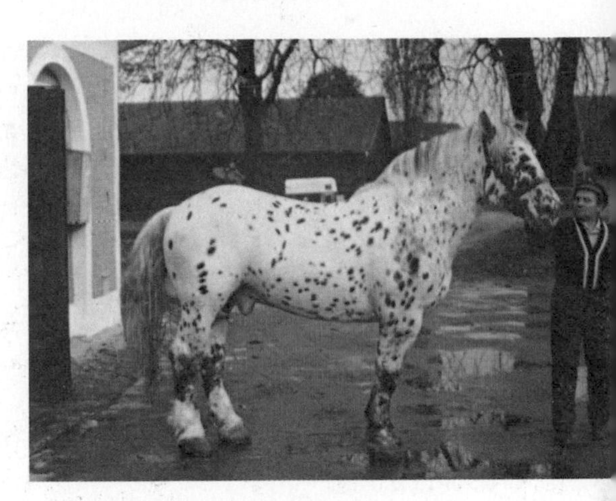

161. Color: *black patterned leopard*
Breed: Noriker
Owner: Austrian government
Photographer: D. P. Sponenberg
Genotype: *aaB_CCddE_Lp_*

162. Color: *patterned leopard sorrel alazán* with extensive *flecked blanket (leopard spots* have *roan* edges and dark centers); blaze; three-quarter stocking on left rear
Breed: Appaloosa
Name: The T.Q. Dude
Owner: T. D. Harper
Photographer: B. Beaver
Genotype: *____CCddeeffstystyLp_* plus *blanket*

164. Color: *unpatterned leopard chestnut tostado*; blaze; stockings on both rear legs
Breed: Appaloosa
Name: Shalakos Plaudit
Owner: Gene Palmer
Photographer: Gene Palmer
Genotype: *____CCddeeffstystyLp_*

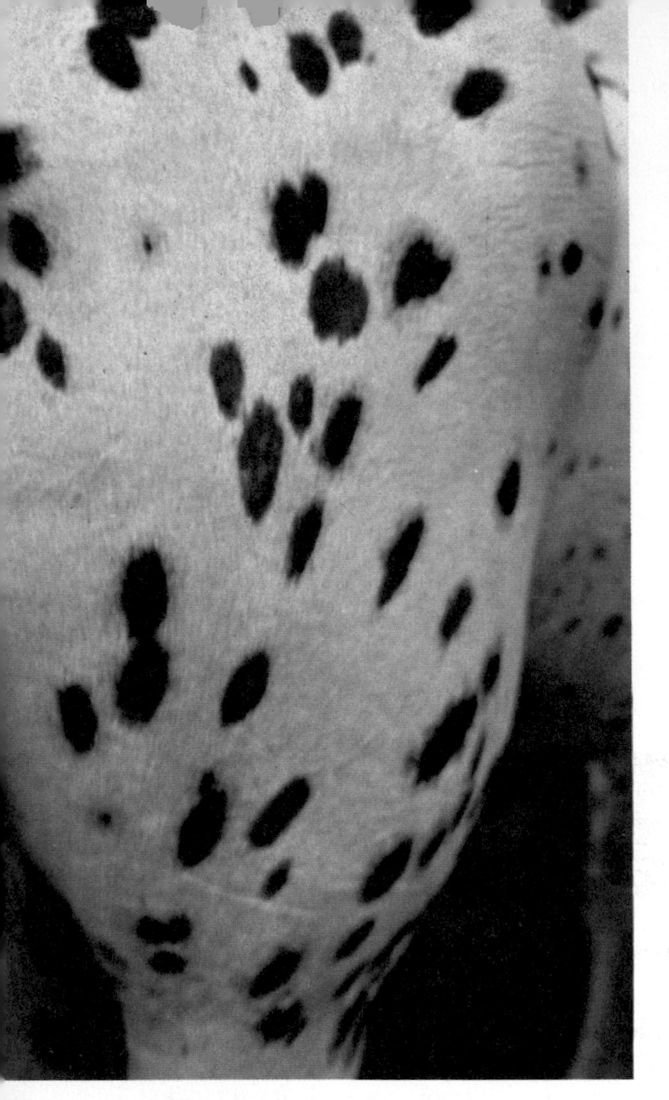

163. Color: *patterned leopard chestnut*
Name: Irma
Photographer: J. K. Wiersema
Genotype: *__B_CCddeeSty_Lp_*

165. Color: *few-spot leopard*
Breed: Appaloosa
Photographer: B. Beaver
Genotype: color? plus *Lp_*

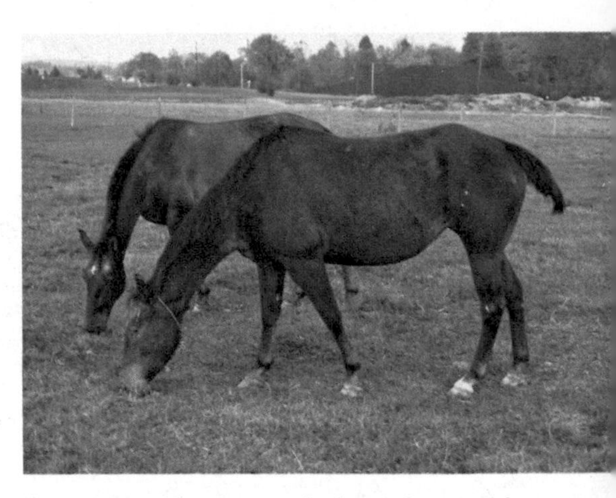

166. Color: *chestnut tostado, leopard spots* with no white; pastern on left rear leg
Breed: grade
Photographer: D. P. Sponenberg
Genotype: *__B_CCddeeffstystyLp_*

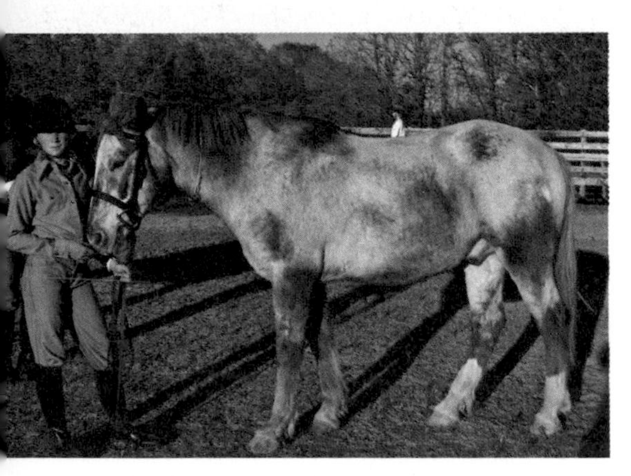

167. Color: *sorrel alazán varnish roan*; star with connected strip; three-quarter stockings on both rear legs
Breed: Appaloosa
Owner: Sid Loveless
Photographer: D. P. Sponenberg
Genotype: *____CCddeef_stysty* plus *varnish roan*

168. Color: *chestnut* (?) *varnish roan*; blaze; stockings on both rear legs and right front, sock on left front
Breed: Appaloosa
Owner: Virginia Polytechnic Institute and State University
Photographer: D. P. Sponenberg
Genotype: color? plus *varnish roan*

169. Color: In front: *snowflake mahogany bay*; coronet on right rear, fetlocks on right front and left rear. Behind: *sabino grey*, born ?
Breed: Spanish Mustang
Name: Whiskey Pete
Owner: Pete Hansen
Photographer: Marye Ann Thompson
Genotype: *A_B_CCddE_Sty_* plus *snowflake*

171. Color: *mottled bay*
Breed: Appaloosa
Name: City Lights
Owner: Larry and Joyce Slusher
Photographer: D. P. Sponenberg
Genotype: *A__B_CCddE_* plus *mottled*

173. Color: *black varnish roan* with *solid blanket* and *dark spots*
Breed: Appaloosa
Photographer: B. Beaver
Genotype: *aaB_CCddE_* plus *blanket* plus *dark spots* plus *varnish roan*

170. Color: *chestnut speckled*; stripe; fetlocks on both rear legs
Breed: American Indian Horse
Name: Orphan
Owner: Buddy and Leana Rideout
Photographer: D. P. Sponenberg
Genotype: *__B_CCddeeffStysty* plus *snowflake*

172. Color: *mottled bay*; socks on both hind legs
Breed: Appaloosa
Name: City Lights
Owner: Larry and Joyce Slusher
Photographer: D. P. Sponenberg
Genotype: *A_B_CCddE_* plus *mottled*

174. Color: *snowflake chestnut ruano varnish roan* with *solid blanket* and *dark spots*; stripe on face; fetlock on right front, three-quarter stockings on both rear legs
Breed: Appaloosa
Photographer: B. Beaver
Genotype: *__B_CCDdeeffSty_* plus *blanket* plus *dark spots* plus *snowflake* plus *varnish roan*

175. Color: *speckled black* (?) *with dark spots*
Breed: Appaloosa
Photographer: B. Beaver
Genotype: color? plus *dark spots* plus *snowflake*

176. Color: (?) *varnish roan* with *blanket*
Breed: Appaloosa
Photographer: D. P. Sponenberg
Genotype: color? plus *varnish roan* plus *blanket*

177. Color: *blue roan (black roan)* with *solid blanket* and *dark spots*
Breed: Spanish Mustang
Name: Majuba
Owner: Emmet Brislawn
Genotype: *aaB_CCddE_Rr*

178. Color: *mahogany bay varnish roan*
Breed: American Indian Horse
Name: Fox
Owner: Nanci Falley
Photographer: Nanci Falley
Genotype: *A_B_CCddE_Sty_*

179. Color: *white*
Breed: Tennessee Walking Horse
Name: Regal
Owner: Raymond Hale
Photographer: Doug Gregg
Genotype: color? plus *Ww*

180. Color: *white*
Breed: American Albino
Name: Missi Blue Taf
Owner: Karen Wales-Patton
Photographer: Alda Buresh
Genotype: color? plus *Ww*

183. Color: *black*; odd facial mark
Breed: Spanish Mustang
Owner: Kim Kingsley
Photographer: Kim Kingsley
Genotype: aaB_CCddE_

184. Color: *chestnut alazán* with freeze brand; star; coronet on right front, fetlock on right rear, white spot on front of left rear
Owner: Virginia Polytechnic Institute and State University
Photographer: D. P. Sponenberg
Genotype: __B_CCddeef_Sty_

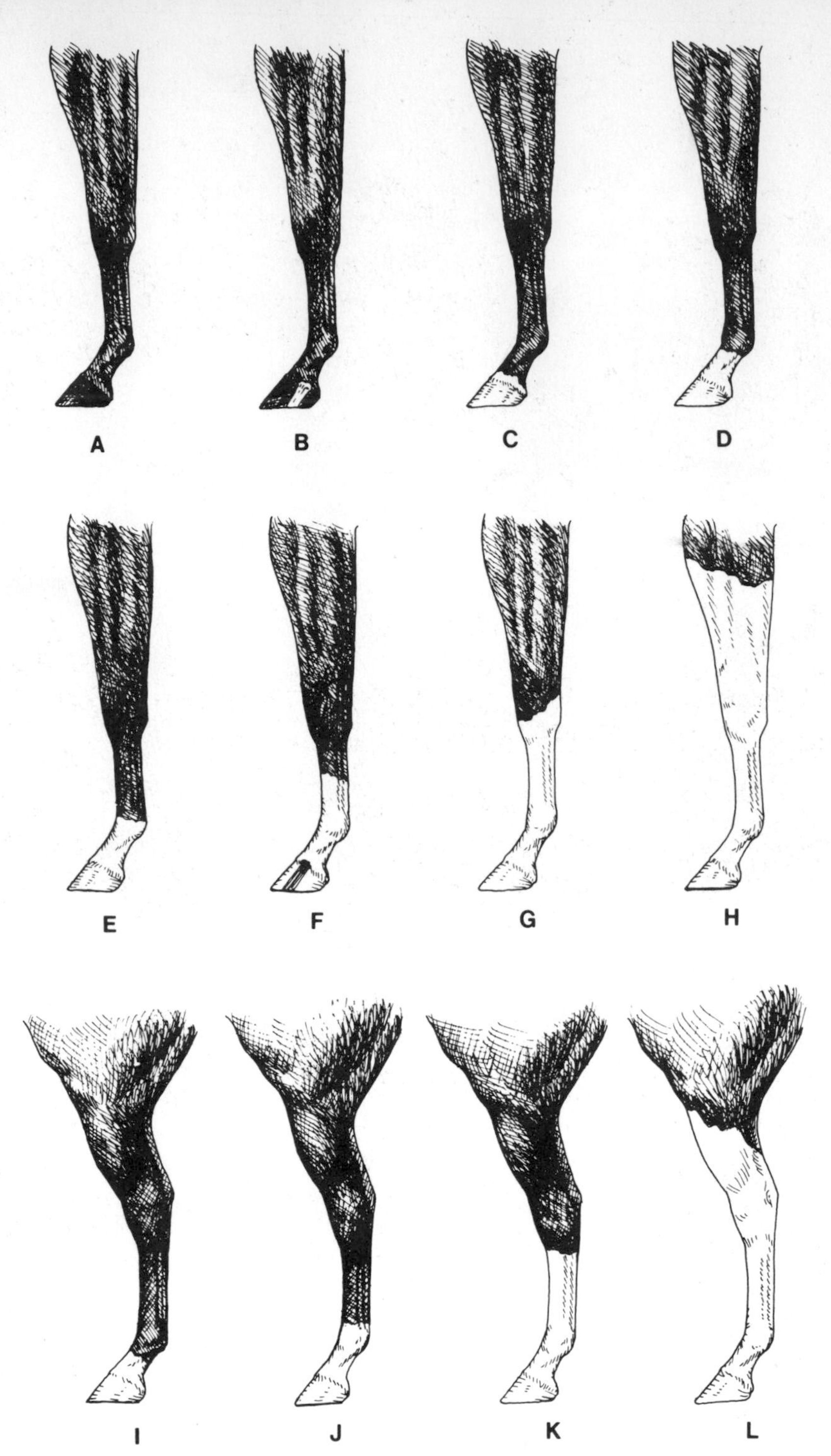

181. *White marks* on horse legs: *A*, unmarked left forelimb; *B*, *white mark* on the outside of the left forelimb; *C*, *coronet* on the left forelimb; *D*, *pastern* on the left forelimb; *E*, *fetlock* on the left forelimb; *F*, *sock* with *ermine mark* on the left forelimb; *G*, *stocking* on the left forelimb; *H*, *extensive stocking* on the left forelimb; *I*, *half pastern* on the left rear limb; *J*, *high fetlock* on the left rear limb; *K*, *three-quarter stocking* on the left rear limb; *L*, *extensive stocking* on the left rear limb

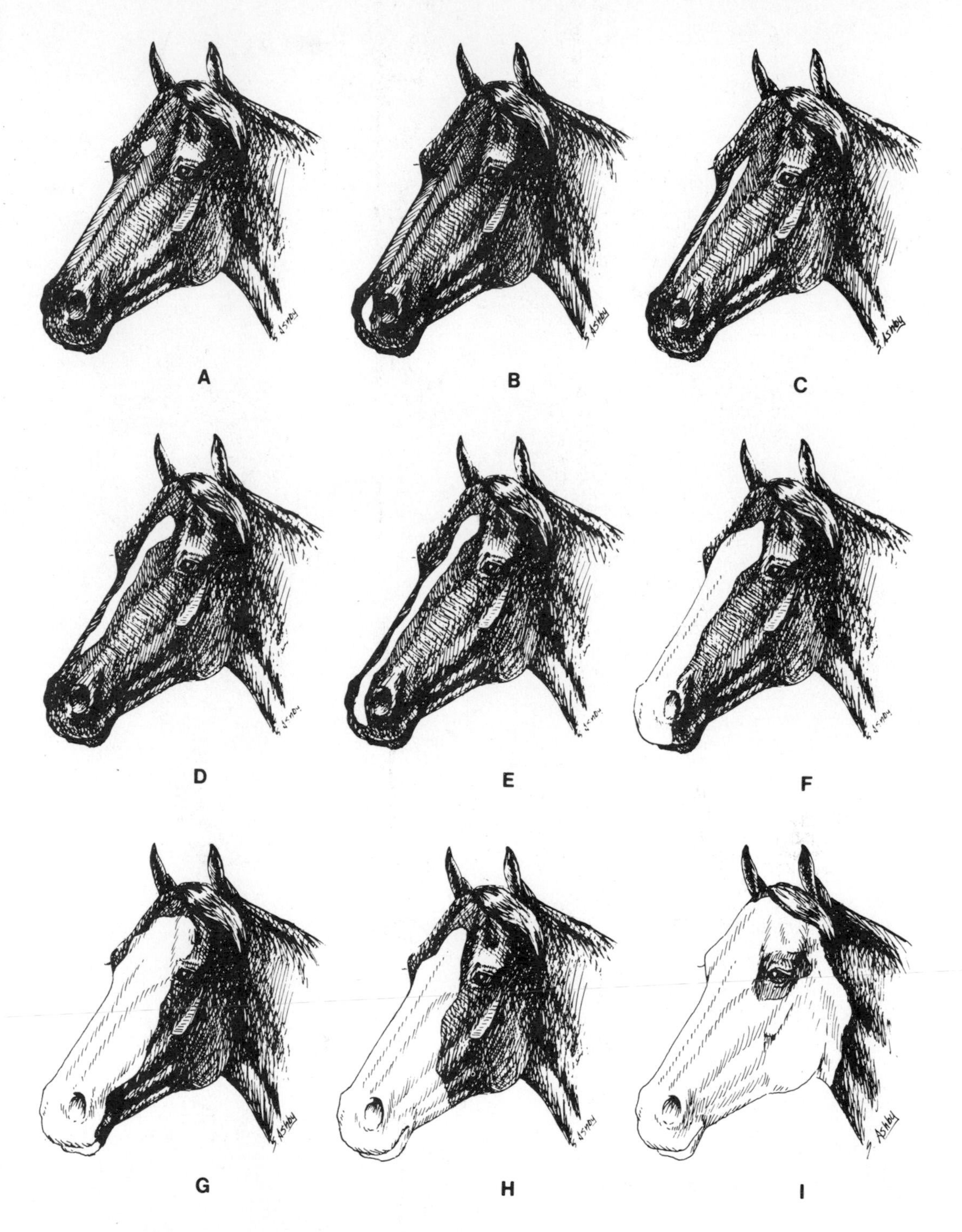

182. *White marks* on horse faces: *A, star; B, snip; C, strip; D, star* and connected *strip; E, stripe (connected star, strip,* and *snip*); *F, blaze; G, bald; H, apron face; I, bonnet*

185. Color: *silver dappled bay*
Breed: Icelandic Horse
Photographer: J. K. Wiersema
Genotype: *A_B_CCddE_Z_*

186. Color: *silver dappled bay* dam and foal
Breed: Welsh Pony
Owner: Trish Cummings
Photographer: D. P. Sponenberg
Genotype: *A_B_CCddE_Z_*

187. Color: *dappled grey* born *black* and *dappled grey* born *sorrel* (two-year-old)
Breed: Quarter Horse
Photographer: B. Beaver
Genotype: born *black*, *aaB_CCddE_G_*; born *sorrel*, *____CCddeeffstystyG_*

some) is recessive to bay *A* but dominant to *black a* and is postulated by some to cause *seal brown*. The actual factor causing *seal brown*, though, is probably not located at the *A* locus, as is explained under the discussion of *pangaré*.

The *chestnuts* resulting from *aabb* would best be called *chocolate chestnuts* rather than *liver chestnut* or *chestnut*, as *chocolate chestnut* would give an indication of genotype. Unfortunately the reality is not that simple, since the *aabb* horses run a spectrum from very dark chocolate, easily confused with *liver chestnut*, to a lighter *chestnut* shade. The manes and tails on some *chocolate-chestnuts* also fade to a redder brown, making point color determination a little tricky also. The skin on *bb* animals is a brownish pink color, in contrast to the black skin of *B* horses. The most sensible approach is to retain the *liver chestnut* or *chestnut* designation unless a horse is obviously a *chocolate chestnut*. The *bb* genotypes are more easily identified when they occur in conjunction with other genes listed below, which is, no doubt, why the existence of *bb* is not well documented.

Black, as in *aaB_*, should be uniformly recessive to *bay*, and *black* horses bred together should always produce *blacks*. In a number of instances this does not hold true. *Black* Clydesdales, for instance, routinely produce *bay* and *brown* foals as well as the expected blacks. It is postulated that a dominant allele is reponsible for this, and it is probably at the *E* locus. This allele is called E^d, and the following genotypes would be *black*: $A\_B\_E^d\_$ and $aaB\_E^d\_$. *Jet blacks* are reputed by some to be due to the E^d allele. If the *B_* is replaced by *bb*, the result ($A\_bbE^d\_$ and aab-$bE^d\_$) would be *chocolate chestnut*, since *bb* horses can only form brown eumelanin. Some people doubt the existence of the E^d allele, but it does help to explain the *bay* and *brown* offspring produced in crosses of *black* horses. All horses with phaeomelanic points (red/flaxen) have body colors that are dictated by the recessive allele, the *ee* genotype. These *ee* animals all have basically some shade of phaeomelanin (red/yellow).

The smutty or sooty effect is a result either of eumelanic (black/brown) hairs mixed in among the base color of the body or of a eumelanic tip on phaeomelanic hairs. These two possibilities (black hairs vs. black tips) are probably genetically distinct but result in similar colors. The smutty effect seems to be dominant but is poorly documented. The gene symbol *Sty* (for smutty) is used for both effects to aid the discussion in this book. The actual situation is probably more complex. Eumelanic hairs are more common than just eumelanic tips and are usually more abundant dorsally on the horse, causing the top of the body (croup and withers) to be darker than the lower portions. This effect is very noticeable when *B_* allows black eumelanin, but it could be missed visually in *bb* horses, because the chocolate hairs mixed with reds or chocolates would be barely discernible. The smutty effect causes *sorrel* to be *chestnut* (*liver chestnut*, if extreme), *bay* to be *mahogany bay*, *slate grullo* to be *lobo dun*, *zebra dun* to be *coyote dun*, and *palomino* to be *smutty palomino*. The recessive, nonsmutty *sty*, symbolizes the clear, pure colors.

The discussion has so far taken care of those colors that are usually referred to as hard colors: *black*, *brown*, *bay*, *sorrel*, and *chestnut*. The other colors are lumped together under the general description of *dun* and are caused by alleles at two loci: *D* and *C*.

The dun factor acts on the body color, but not the point color, to cause dilution of black to slate blue, chocolate brown to a lighter buff brown, and red to light red or yellow. The dun factor, *D*, is dominant to the nondun allele, *d*. The dun allele does not affect points or the black hairs caused by the *Sty* allele. *Black* with the dun factor (*aaB_D_E* or $\_\_B\_D\_E^{d}\_$) is a shade in the grullo group (*slate grullo*, *olive grullo*, or *lobo dun* if smutty). *Bay* is diluted to *zebra dun* (*A_B_D_E_*) or *coyote dun* if smutty. *Chestnut/sorrel* is diluted to one of the *red dun* shades (*____D_ee*) or to *yellow dun* (*A_bbD_E_*). Genetically, some *muddy duns* (aabbD_E_) are *chocolate grullos*. Other *muddy duns* are the results of *D* on very dark-pointed *ee* horses. The primitive markings seen on the horses of these colors are sometimes attributed to the *D* allele itself, but since these marks appear on some of the nondilute colors (bay), they are probably caused by a separate factor. The *D* allele occurs in the North American Spanish Horse, the Quarter Horse, and the Icelandic Horse as well as in all horses of the Norwegian Fjord and Tarpan breeds. It does not occur in the Thoroughbred or the Arabian.

The cremello allele, c^{cr}, is incompletely recessive to nondilute *C*, but it only acts to dilute phaeomelanic red to yellow, having little effect on eumelanic black/brown on the body or the points. The heterozygote (Cc^{cr}) is diluted a little: *bay* ($A\_B\_Cc^{cr}ddE\_$) becomes *buckskin,* and *chestnut/sorrel* becomes *palomino* ($\_\_\_\_Cc^{cr}ddee$) or *smutty palomino* if *Sty* is present and *yellow dun* ($A\_bbCc^{cr}ddE$) or *claybank dun* if the mane and tail remain red or dark brown. The *F* locus may determine mane and tail color on these dilute colors, but the c^{cr} allele itself may be responsible for light manes and tails regardless of the allele present at *F. Black* and the *chocolate chestnuts* are hardly diluted at all. The *black* can become *smoky black,* but many black horses known to carry a c^{cr} gene are dark black with no visible effect of the c^{cr}. The *chocolate chestnuts* (aabb) can likewise be lightened up to *smoky chestnuts.* The effect on *chocolate* is greater than that on *black.* Some of these, as in Figs. 53 and 54, would not be classified as *chestnuts* but as *lilac duns.* The homozygotes ($c^{cr}c^{cr}$) are lightened to a cream color with blue eyes. The points on these $c^{cr}c^{cr}$ horses are also dilute, with black points being pale red or pale blue and red/flaxen points being off-white. The homozygous bays are *perlino,* and the homozygous chestnuts/sorrels are *cremello.* The homozygous effect on *black* is not documented, but it is probably a "*silver smoky*" with blue eyes. Likewise, the homozygous effect on *chocolate chestnut* is not documented. On *grullo* the homozygous effect is *silver grullo* ($aaB\_c^{cr}c^{cr}D\_E\_$).

The *pangaré* effect, *P,* is dominant to its absence, *p.* This effect causes light phaeomelanic areas on the muzzle, over the eyes, in the flanks, and on the inside of the legs. It can be present on any color but does not change the name of the color except on *black,* which is changed to *seal brown* ($aaB\_ccddE\_P\_$). *Pangaré* is present on all *blond sorrels* and on some darker *sorrels* as well.

The silver dapple allele, *Z,* not only dilutes eumelanic points but also lightens any eumelanin on the body. As such it has profound effects on *black,* causing it to be modified to a sepia brown with lighter dapples. It does not effect phaeomelanin, so even though it will dilute the points of a *bay* horse, it will not affect the body color.

The number of loci affecting colors (at least the loci that are fairly well documented) is nine: *A, B, C, D, E, F, P, Sty,* and *Z.* The way they interact can make it difficult to predict colors that certain crosses will produce or to assign genotypes on the basis of visual appraisal alone. The group with black points is fairly straightforward, since they are all modifications of *B_*, and only the *A, C, D, P*, and *Sty* loci can have an effect on the body color.

The colors with nonblack points are the most confusing, because they are all lumped together in the visual classification but can be caused by the *bb*, *ee*, or *Z_* genotypes. These shades, if undiluted by c^{cr} or *D*, are all called *sorrel*, *chestnut*, or *liver chestnut*, depending on shade. The *aabb* and *eeSty_* *liver chestnuts* may look very similar visually if the *eeSty_* has a very dark mane and tail. *Silver dappled blacks* (*aaB_CCddE_Z_*) usually are fairly easily identified, but if the dappling is not pronounced, they may resemble flaxen-pointed *chestnuts* or *liver chestnuts* (*____CCddeeffSty_*). *Silver dappled bays* (A*_B_CCddE_Z_*) could be confused with flaxen-pointed *sorrels* (*____CCddeeffstysty*). The points on the *silver dapples* usually have a little of the darker color in them, which helps in distinguishing them from the phaeomelanic red/flaxen points. The presence of the *silver dappled* "*sorrels*" is evidently more widespread in Europe than in the United States and can result in *bays* being born to *sorrel* parents: (*AABBCCddEE__Zz*) × (*AABBCCddeeffzz*) = (*A_B_CCddE_ffzz*), which is *bay*. As mentioned before, though, most breeds in the United States lack the *b* allele and the *Z* allele, and so most *sorrel/chestnut/liver chestnut* horses are caused by the *ee* genotype.

The *dun* and *cremello* loci can also cause confusion, except that the *D* allele is usually associated with primitive markings and the c^{cr} allele is usually not. *Red duns*, *zebra duns* and *grullos* result from the *D* allele, and *palominos* and *buckskins* are caused by the c^{cr} allele. *Yellow dun* body color can be a result of either allele, with the same color resulting but with *D* present in the linebacked animals. The effect of combining the dun allele with the heterozygous cremello allele, $Cc^{cr}D$_, on horses is not additive: the horse is no lighter than it would be with a single effect alone. *Black* with $Cc^{cr}D$ would be *grullo*, *bay* would be *zebra dun*, and *chestnut/sorrel* would be *palomino* (with a back stripe).

As a final word on the genetics of color, it should be noted that in most breeds the only loci that determine color are *A*, *E*, *F*, *P*, and *Sty*. Although the *B*, *C*, *D*, and *Z* loci can become important, they are usually rare, and the exceptions have already been noted.

Genetics of Patterns of White

The inheritance of the patterns of white is easier to understand than the inheritance of colors, since each independent locus causes a distinct pattern of white. Most of these patterns, especially the more common ones, are not genetically complex.

Greying, *G*, is dominant to nongreying, *g*. The homozygote is reputed to become white faster than the heterozygote, but this is not certain. The rate at which a horse turns light may well be governed by independent modifiers (for example, the two-year-old *grey* Quarter Horses in Fig. 187 are greying at different rates). In breeds in which dark *grey* is a popular color, selection for slow greying has resulted in horses that stay dark to medium *grey* for much longer than horses of breeds in which no such selection has taken place. Percheron horses in the United States are an example of a breed that has been selectively bred to turn white slowly.

The reason for the mane and tail turning light faster than the body on some horses and not on others has not yet been attributed to a genetic cause. The genetics of the two types of *grey* (one turning completely white and the other stabilizing at the darker mane, legs, and head) also has not been studied in detail. Whether this difference is genetic or not is unknown.

Roan (*R*) is usually cited as being a dominant lethal to non*roan* (*r*), meaning that the heterozygotes are *roan* and homozygotes do not exist because they die during development. Most *roans* are heterozygous and throw solid-colored foals in addition to *roans*, but homozygotes are known in the Dutch and Brabant draft horses. Rare exceptions, however, probably prove the rule, and *roan* should be considered as a dominant lethal trait.

The *rabicano* pattern is probably dominant (*Rb*), occurring in certain families of various breeds of horses (Quarter Horse, Noriker, Arabian, Welsh, Hackney).

The *frosty* pattern has not been characterized genetically.

Tobiano (*T*) is dominant to nonspotted (*t*). Both homozygotes and heterozygotes occur, and both are spotted. The amount of white is probably governed by independent modifiers, and breeders can select for either extensive or little white.

Overo (*o*) is considered recessive to nonspotted (*O*), but the actual situation is probably more complex than this. Confusion of *sabino* with *overo* probably contributes to the confusion of the inheritance of *overo*. Breeding *overo* to *overo* sometimes results in solid foals, and this should not happen if the pattern is a simple recessive. One horse with four colored feet and a bald face produced foals as if he were an *overo* genetically, so he may have been an example of an *overo* that had such little spotting that it was only present as extreme white on the face. By visual appraisal, he would have been classified as nonspotted. This phenomenon probably does not explain all the solid foals born to *overo* breedings, since solid foals happen rather frequently. At the other extreme are the *overos* that are mostly or all white. Many of these die within a few days of birth because of malfunction of the colon, although some white foals born of *overo* parents are normal and survive.

Sabino (*Sb*) has not been extensively studied but is probably dominant to nonspotted (*sb*). The minimum expression seems to be high white stockings, perhaps not on all four feet, and extreme facial white. These horses would be classified as non*sabino* and therefore make genetic studies difficult. The heterozygotes may be less extensively white than the homozygotes. Some white foals are born of *sabino* × *sabino* matings and in turn produce *sabinos* when bred to solid horses. These white foals develop normally in contrast to the white foals of the *overo* breedings. In the Clydesdale breed *sabino* is erroneously called "*roan*," and *sabino* × non*sabino* breedings produce a high proportion of the desired high white stockings and blazes on the face. When *sabinos* are bred to *sabinos*, though, a high percentage of the foals are excessively white, and these tend to have the *medicine hat* pattern that is popular in some breeds of horses such as the North American Spanish.

The *splashed white* pattern (*spl*) is recessive to nonspotted (*Spl*). It is limited to certain breeds of European horses, including the Finnish Draft Horse, the Welsh Pony, and other northern European breeds.

The genetic situation concerning the symmetrical patterns of white (*appaloosa*) is confused at best. Many hypotheses have been put forth, but none seems to explain all cases. Several independently inherited patterns may be present and may be responsible for the difficulty of determining the inheritance. *Blankets* occur as *roan*, *solid*, or *flecked*, and whether these three represent different manifestations of the same genetic effect or are genetically different is unknown. Extensively blanketed horses may look like *leopards* if the dark spots are present, but these spots may be an independent trait. Some *leopards* that are mostly white may not be related to the *blanket* patterns at all. The two types of *leopards* (*patterned* and *unpatterned*) may also be genetically distinct. *Varnish roan* and *snowflake/speckled* are both probably independent of the *blanket* patterns.

Evidence in the Noriker breed indicates that *blankets* and the *leopard pattern* are different extremes of the same pattern. In the Noriker, homozygotes (*LpLp*) are the *few-spot leopard* pattern, and heterozygotes (*Lplp*), an incomplete dominant, are the *leopard pattern*. In this breed the situation is fairly simple, as other patterns do not occur. Even so, *leopard* foals from non-*leopard* breedings indicate that the genetic situation may be more complex than most of the evidence indicates.

Most Appaloosa breeding revolves around the *blanket* patterns and depends on the fact that when *blanketed* or *leopard* horses are bred to solid-colored horses of breeds not having the patterns, most resulting foals will be patterned. This indicates at least a tendency for the patterns to be dominant. However, extensively marked horses such as *leopards* have cropped out from certain lines of Quarter Horses and Andalusians after generations of solid × solid breedings. This happens frequently enough to discount errors of parentage, especially since such marked Quarter Horses are not eligible for registration as Quarter Horses. These patterned individuals indicate some recessive tendencies for the patterns. All of this causes considerable confusion, some of which, no doubt, results from difficulty in accurately determining the patterns present and some from not really knowing how many independent patterns are in existence. *Varnish roan* and *snowflake/speckled* may be easier to identify but have not been extensively studied as to genetic mechanisms. Some studies on the genetics of the Appaloosa are available, and the interested reader is urged to look at several of these for pertinent theories of genetic mechanisms.

White foals can result from several mechanisms. One type

of white (*W*) is a dominant lethal, so, as in the *roan*, only heterozygotes exist. Other types result from *overo* matings, and these usually die. *Sabinos* sometimes produce *whites*, and so do the symmetrical patterns of white (specifically the *blanket* and *leopard* patterns). *White* horses resulting from the breeding of two solid-colored horses occur with enough frequency to make one wonder if the above mechanisms are the only way to get *white* foals. Two breeds in which this *white* foaling occurs, the Thoroughbred and the Brabant, do not have many *white* horses, so errors of parentage are unlikely The different types of *white* horses cannot be distinguished from one another visually, but usually they can be determined by knowing the appearance of the parents or the offspring of the horse under consideration. The "whites" that occur from breeding palominos together are actually *cremellos* and are a result of the $c^{cr}c^{cr}$ genotype.

Minor white marks on the face and legs do not behave consistently as either dominant or recessive genetic effects. They may be governed by quantitative inheritance, but this is uncertain. Selection for or against the facial and leg white is generally quite effective in breeds like the Cleveland Bay. Since white has been selected against, Cleveland Bays have very little. Other breeds, such as the Clydesdale and some breeds of gaited horses, select for white markings and generally end up with four white legs and facial white. In the Clydesdale this patterning is complicated by the *sabino* gene, but in other breeds it is not. Another interesting fact is that experimentally induced identical twins, carried in different mares, can have very different markings of white. This indicates that another major factor, unrelated to genetics, is determining the extensiveness of the white marks.

The following list demonstrates that a given color can have more than one genotype. *Coyote dun*, for example, can be *A_B_CCD_E_Sty_* or $A\_B\_Cc^{cr}D\_E\_Sty\_$. Other examples are obvious from the list and serve to indicate how complex breeding for specific colors can be.

The loci:

- *A* restricts eumelanin to points (*bay*)
- *a* eumelanin over entire body, uniform black/brown
- *B* black eumelanin

b	brown eumelanin
C	non-dilute
c^{cr}	*cremello* dilution—dilutes phaeomelanin markedly, eumelanin a little
D	*dun* dilution—dilutes both eumelanin and phaeomelanin
d	non-dilute
E^{d}	dominate black, epistatic to *A* locus
E	allows eumelanic points
e	phaeomelanic points, epistatic to *A* locus
F	red points on *ee* horses (*tostado* or *alazán*)
f	flaxen points on *ee* horses (*ruano*)
P	*pangaré* (light ventral areas, muzzle, over eyes)
p	non*pangaré*
Sty	*smutty*—black mixed into body coat
sty	clear, non*smutty* body color
Z	*silver dapple*—dilutes eumelanin only
z	non-dilute
G	*grey*
g	non*grey*
R	*roan*
r	non*roan*
T	*tobiano* spotting
t	nonspotted
O	nonspotted
o	*overo* spotting
Sb	*sabino* spotting
sb	nonspotted
Spl	nonspotted
spl	*splashed white*
Rb	*rabicano*
rb	non*rabicano*
Lp	*leopard*
lp	non*leopard*
W	*white*
w	non*white*

Unidentified genetic mechanisms:
blanket (solid, roan, flecked)
snowflake/speckled

varnish roan
facial white marks
leg white marks
frosty

Genotypes of the Colors

Determining the genotype of given horse by visual appearance can be difficult, since some visually discernible differences in color, such as *slate grullos* and *olive grullos*, *bay* and *brown*, the non*ruano* points on *chestnuts* and *sorrels* (meaning the brown, red, and darker flaxen points), and *claybank dun* and *palomino*, have not yet been attributed to genetic differences.

The opposite situation of charting all possible genotypes and then assigning colors to the various combinations can also be confusing. Some combinations would be quite rare, and a theoretical guess at the resulting color is all that would be possible. Occasionally, though, one does come upon a truly rare combination, and being familiar with the actions of the various genes can help in understanding the color in question.

In this chapter the loci that determine color—*A*, *B*, *C*, *D*, *E*, *F*, *Sty*, *Z*, and *P*—are considered. All combinations of seven of the loci will be considered: *A*, *B*, *C*, *D*, *E*, *F*, and *Sty*. Since *Z* and *P* only interact to change identification and color names with a few combinations, they do not need to be categorized for all combinations. The *B* list is somewhat abbreviated, since loci *A* and *B* have no effect on *eeStysty* horses (except to cause subtle changes of skin color). *F* has no effect on *E* or E^d horses, and *Sty* probably has little effect on *bb* horses.

Also in this chapter a shorthand notation is used. For example, *A* will stand for genetypes *AA* or *Aa*, since they are visually identical, and *a* will stand for *aa*. The locus *C* must be represented by two letters, since *CC*, Cc^{cr}, and $c^{cr}c^{cr}$ all are visually different. The order of the loci is *ABCCDEFSty*. Dashes will be used to represent entire loci that have no effect in a given combination. For example, in *--CCDefsty* the *A* and *B* loci are unimportant, as any combination of these would still result in a *sorrel ruano*, although skin color of *bb* would be lighter than that of *B*. (Note that in the previous chapter, as in the figure captions, this combination would have been given as *____CCD_eeffstysty* to indicate the full arrangement of the genes.)

The list will start at the completely dominant situation and go towards the completely recessive situation. The most likely color will be presented, with other possibilities in parentheses. Question marks indicate the hypothetical colors that result from some combinations. This list is also summarized in Table 1.

ABCCDE–Sty	*Coyote dun*. The points are black (*B*, *E*), the body is yellow (*A*, *D*) and primitive marks are present (*D*). In addition, the body coat is mixed with black.
ABCCDE–sty	*Zebra dun*. This combination lacks the black in the body coat (*sty*).
$-BCCDE^{d}-Sty$	*Lobo dun* (*smutty olive grullo*). The slate body results from the combination DE^{d}.
$-BCCDE^{d}-sty$	*Slate grullo* (olive grullo).
–BCCDeFSty	*Dark red dun* (?). This is theoretical, as *Sty* should cause black throughout the coat. This combination may not be smutty, however, and may be normal *red dun*.
––CCDefsty	*Red dun* (*orange dun*; *apricot dun*).
–BCCDefSty	*Dark apricot dun* (?). Again, the smuttiness is theoretical. The light points from *f* would probably lighten the color enough to be *apricot dun* instead of *red* or *orange dun*.
––CCDefsty	*Apricot dun*.
ABCCdE–Sty	*Mahogany bay* (*brown*).
ABCCdE–sty	*Bay* (*red*, *blood*, or *sandy*; *brown*).
$-BCCdE^{d}--$	*Jet black*.
–BCCdeFSty	*Chestnut* (*liver chestnut*), either *tostado* or *alazán*.
––CCdeFsty	*Sorrel*, either *tostado* or *alazán*.
–BCCdefSty	*Chestnut* (*liver chestnut*), *ruano*.
––CCdefsty	*Sorrel*, *ruano*.
$ABCc^{cr}DE-Sty$	*Coyote dun*. The *CCD coyote duns* might be expected to have darker heads than the

$Cc^{cr}D$ *coyote duns*. The *D* allele acting alone generally allows a head that is somewhat darker than the body. The C^{cr} allele acts to lighten the head along with the body to the same color. The primitive marks would still be present from the *D* allele.

$ABCc^{cr}DE{-}sty$ — *Zebra dun*. This combination may have a lighter head than that of *CCD*.

$-BCc^{cr}DE^{d}{-}Sty$ — *Lobo dun* (*smutty olive grullo*). The lighter head is not present, as c^{cr} does not act much on black (E^{d}).

$-BCc^{cr}DE^{d}{-}sty$ — *Slate grullo* (*olive grullo*). The same remarks about the lighter head of the *lobo dun* would hold true here. *Silver grullos* ($c^{cr}c^{cr}D$) do have darker heads, which indicates that the head color of *black* or *grullo* horses is not lightened with c^{cr}.

$-BCc^{cr}DeFSty$ — *Smutty linebacked palomino* (*smutty linebacked claybank dun*?). The c^{cr} allele may act on horses that are *eF* or *ef* to produce an identical color. *Claybank duns* do occur, though, and are difficult to explain unless some animals that are $Cc^{cr}ef$ retain red or brownish manes and tails, with little of the light *palomino* color in them.

$--Cc^{cr}DeFsty$ — *Linebacked palomino* (*linebacked claybank dun*).

$-BCc^{cr}DefSty$ — *Smutty linebacked palomino*.

$--Cc^{cr}Defsty$ — *Linebacked palomino*.

$ABCc^{cr}dE{-}Sty$ — *Dark buckskin*. No primitive marks because *D* is not present.

$ABCc^{cr}dE{-}sty$ — *Buckskin*.

$-BCc^{cr}dE^{d}--$ — *Smoky black*. Some are not even lightened to *smoky black* and remain black.

$-BCc^{cr}deFSty$ — *Smutty palomino* (*smutty claybank dun*).

$--Cc^{cr}deFsty$ — *Palomino* (*claybank dun*).

$-BCc^{cr}defSty$ — *Smutty palomino*.

$--Cc^{cr}defsty$ — *Palomino*.

TABLE 1. Interactions of the color genes. The loci A, B, E, Sty, F, and P have complex interactions which determine point and body color. These actions are shown in Part I of the chart as each locus is added to the basic situation. The last line (P/p) of Part I reveals the final combination of all these effects. Each line of Part II is separate and unrelated to other lines of Part II but shows the interactions of the last line of Part I (the P/p line) with various combinations of the C, D, and Z loci.

I

<table>
<tr><td>B/b
$E^d/E/e$</td><td colspan="4">BE^d
jet black</td><td colspan="6">BE black points</td><td colspan="2">bE brown points</td><td colspan="4">$-e$ red/flaxen points</td></tr>
<tr><td>A/a</td><td colspan="4">no
effect</td><td colspan="4">a
black</td><td colspan="2">A
bay/brown</td><td>a
chocolate
chestnut</td><td>A
sorrel
tostado</td><td colspan="4">no
effect</td></tr>
<tr><td>Sty/sty</td><td colspan="2">Sty
jet
black</td><td colspan="2">sty
jet
black</td><td colspan="2">Sty
black</td><td colspan="2">sty
black</td><td>Sty
Mahogany
bay
dark brown</td><td>sty
other
bays
lt. brown</td><td colspan="2">no
effect</td><td colspan="2">Sty
chestnut/
liver chestnut</td><td colspan="2">sty
sorrel</td></tr>
<tr><td>F/f</td><td colspan="4">no
effect</td><td colspan="6">no
effect</td><td colspan="2">no
effect</td><td>F
tostado/
alazán</td><td>f
ruano</td><td>F
tostado/
alazán</td><td>f
ruano</td></tr>
<tr><td>P/p</td><td>pp
jet
black</td><td colspan="2">P
seal
brown</td><td>pp
jet
black</td><td>pp
black</td><td colspan="2">P
seal
brown</td><td>pp
black</td><td colspan="2">does not
change names</td><td colspan="2">does not
change names</td><td colspan="3">does not change names</td><td>P
blond
sorrel</td></tr>
</table>

II

<table>
<tr><td>Cc^{cr}</td><td>smoky</td><td colspan="2">light seal brown</td><td>smoky</td><td>smoky</td><td colspan="2">light seal brown</td><td>smoky</td><td>smutty buckskin</td><td>buckskin</td><td>lilac dun</td><td>yellow dun</td><td>smutty claybank dun/ palomino</td><td>smutty palomino</td><td>claybank dun/ palomino</td><td>palomino</td></tr>
<tr><td>D</td><td colspan="2">lobo dun</td><td colspan="2">slate grullo</td><td colspan="2">lobo dun</td><td colspan="2">slate grullo</td><td>coyote dun</td><td>zebra dun</td><td>muddy dun</td><td>lineback yellow dun</td><td colspan="4">red, orange, or apricot dun</td></tr>
<tr><td>$Cc^{cr}D$</td><td colspan="2">lobo dun</td><td colspan="2">slate grullo</td><td colspan="2">lobo dun</td><td colspan="2">slate grullo</td><td>coyote dun</td><td>zebra dun</td><td>lineback lilac dun</td><td>lineback yellow dun</td><td>lineback smutty claybank/ palomino</td><td>lineback smutty palomino</td><td>lineback claybank/ palomino</td><td>lineback palomino</td></tr>
<tr><td>$c^{cr}c^{cr}$</td><td colspan="4">silver smoky</td><td colspan="4">silver smoky</td><td colspan="2">perlino</td><td>silver lilac dun</td><td>perlino</td><td colspan="4">cremello</td></tr>
<tr><td>$c^{cr}c^{cr}D$</td><td colspan="4">silver grullo</td><td colspan="4">silver grullo</td><td colspan="2">lineback perlino</td><td>silver muddy dun</td><td>lineback perlino</td><td colspan="4">lineback cremello</td></tr>
<tr><td>Z</td><td colspan="4">silver dapple</td><td colspan="4">silver dapple</td><td colspan="2">sorrel ruano</td><td colspan="2">sorrel ruano</td><td colspan="4">no effect</td></tr>
</table>

$A{-}c^{cr}c^{cr}DE{-}{-}$ *Linebacked perlino.*

$ABc^{cr}c^{cr}DE^{d}{-}{-}$ *Silver grullo.* This argues against the C^{cr} allele causing lighter heads on combinations with *black,* since *silver grullos* have dark heads.

${-}{-}c^{cr}c^{cr}De{-}{-}$ *Linebacked cremello.*

$A{-}c^{cr}c^{cr}dE{-}{-}$ *Perlino.*

${-}Bc^{cr}c^{cr}dE^{d}{-}{-}$ *Silver smoky* (?). If this combination occurs, it is quite rare and is probably a light smoky color with blue eyes.

${-}{-}c^{cr}c^{cr}de{-}{-}$ *Cremello.*

$AbCCDE{-}{-}$ *Linebacked yellow dun.* This is essentially a *zebra dun* with chocolate brown replacing the black in the points. The *Sty* allele could possibly cause the color to be darker, but would not be as striking as the *Sty* allele combined with the *B* allele to give the smutty effect.

$AbCCDE^{d}{-}{-}$ *Muddy dun.*

${-}bCCDeFSty$ *Red dun* (*orange dun*). The smuttiness would not be evident, since it would be chocolate hairs in a red coat. The skin would be pinkish brown, not black.

${-}bCCDefSty$ *Apricot dun.*

$AbCCdE{-}{-}$ *Sorrel, tostado.*

${-}bCCdE^{d}{-}{-}$ *Chocolate chestnut.*

${-}bCCdeFSty$ *Sorrel,* either *tostado* or *alazán.*

${-}bCCdefSty$ *Sorrel, ruano.*

$AbCc^{cr}DE{-}{-}$ *Linebacked yellow dun.* Perhaps the head would be lighter than that of *CCD.*

${-}bCc^{cr}DE{-}{-}$ *Muddy dun.*

${-}bCc^{cr}DeFSty$ *Linebacked pink-skinned palomino* (*linebacked claybank dun*).

${-}bCc^{cr}DefSty$ *Linebacked pink-skinned palomino.*

$AbCc^{cr}dE{-}{-}$ *Yellow dun.*

${-}bCc^{cr}dE^{d}{-}{-}$ *Lilac dun.* These are rare and do not look chestnut. The points are chocolate, the body is a dove color, and the eyes are amber. The skin is pinkish brown. The c^{cr} allele has a greater effect on chocolate eumelanin than on black eumelanin. Another designation for this horse could be *smoky chestnut*.

$-bCc^{cr}deFSty$	*Pink-skinned palomino* (*claybank dun*).
$-bCc^{cr}defSty$	*Pink-skinned palomino*.
$-bc^{cr}c^{cr}DE^{d}--$	*Silver muddy dun* (?).
$-bc^{cr}c^{cr}dE^{d}--$	*Silver lilac dun* (?). Probably lighter than *lilac dun*, with blue eyes.
$aBCCD(E,E^{d})-Sty$	*Lobo dun* (*smutty olive dun*).
$aBCCD(E,E^{d})-sty$	*Slate grullo* (*olive grullo*).
$aBCCdE--$	*Black*. As opposed to *jet black* (E^{d}), these are probably black with blacker points. The effect of *Sty* is probably barely visible, if at all.
$aBCc^{cr}D(E,E^{d})-Sty$	*Lobo dun* (*smutty olive grullo*).
$aBCc^{cr}D(E,E^{d})-sty$	*Slate grullo* (*olive grullo*).
$aBCc^{cr}d(E,E^{d})-sty$	*Smoky black*.
$aBc^{cr}c^{cr}D(E,E^{d})--$	*Silver grullo*.
$aBc^{cr}c^{cr}d(E,E^{d})--$	*Silver smoky* (?). Theoretical, but probably lighter than *smoky* and with blue eyes.
$abCCD(E,E^{d})--$	*Muddy dun*.
$abCCd(E,E^{d})--$	*Chocolate chestnut*.
$abCc^{cr}D(E,E^{d})--$	*Muddy dun*.
$abCc^{cr}d(E,E^{d})--$	*Smoky chestnut* (*lilac dun*).
$abc^{cr}c^{cr}D(E,E^{d})--$	*Silver muddy dun* (?).
$abc^{cr}c^{cr}d(E,E^{d})--$	*Silver lilac dun* (?).
P	With any black genotype this gives *seal brown*. When combined with Cc^{cr}, it results in a light *seal brown*.
Z	With any black genotype this results in a *silver dapple*. With other black-pointed colors (*bay*, *brown*) it would mimic *sorrel* if the points were lightened a lot. With *buckskin* it might result in a color near *palomino*.

Afterword

The identification and biology of horse color have been the topics of this book. To a great many cultures the color of horses has been much more than simple identification. Several cultures have held certain colors or patterns in high regard for a variety of reasons. Some have felt that horses of certain colors were more durable than others. The Spanish conquistadors esteemed *chestnuts* as the hardiest and fleetest, as did certain Arab tribes. *Linebacked duns* were preferred by white North Americans in the West, and these are still preferred in many parts of South America. Belgians prefer *roans* (especially *red roans*) as the hardiest and best-built horses. These preferences are usually passed over as superstition by the outside observer, but rarely has anyone proven these assumptions one way or the other. Polish horse breeders, however, have evidence that *chestnut* mares are more successful than *bay* mares over short races but that the reverse is true over long races. *Bay* mares also have longer and more fertile lives than *chestnut* mares in the Polish studs, so different colors may indeed have differences in performance.

Other cultural preferences have been linked more to esthetics than to supposed constitutional traits. Native Americans of the northern tribes preferred spotted horses of all sorts for their decorative qualities. With *medicine hat paints* that preference was coupled with the belief that the rider of such a horse could not be injured in battle. Other purely esthetic preferences include the current popularity of *blond sorrel* in the American Belgian, *red bay* with four white stockings and a blaze in the Clydesdale, and dark colors with no white marks in the Morgan. These fashions can come and go, and the color array of some breeds changes markedly as preferences change.

Most of the colors and patterns of horses are widespread and occur in a number of breeds, so it is difficult to trace where and when they might have originated. The common horse colors of *black, brown, bay, chestnut,* and *sorrel* are so widespread that they probably arose and spread early in the domestication of the horse. The *linebacked duns* are likewise widespread and are common in certain primitive breeds of horse such as Tarpan, Highland Pony, and Norwegian Fjord Horse. It is tempting to suppose that some shades of the *linebacked dun* are ancestral colors of the horse. The *palomino* and other *dun* colors that are not *linebacked* are also widespread, but somewhat less so than the *linebacked* varieties.

The *grey* pattern is common in Arabians and Barbs and in the many breeds they influenced. It is tempting to suppose that the *grey* pattern originated in one of them and spread from there to both Asia and Europe. The *roan* pattern can likewise be traced, in a number of breeds, to the influence of Spanish horses (light horse breeds) and the Brabant or Ardennes (heavy horse breeds). The Ardennes is a very old breed with little outside influence, and it is possible that the *roan* originated in that breed. The *rabicano* pattern is likewise present in a number of breeds that have a Spanish influence.

The origin of the *tobiano* spotting is more obscure, since it occurs in different breeds worldwide. The unusual appearance of any spotting pattern may have made *tobianos* more likely to be traded (or stolen?) and may account for the worldwide occurrence of some of them. *Overo* can be traced to Spanish influence in a great many breeds. *Sabino* can be traced to Spain, from there to the low countries (or vice versa), and from there to the Clydesdale. The *splashed white* pattern occurs in only a few very widely separated breeds such as the Welsh Pony and the Finnish Draft Horse; it is difficult to imagine where it originated. The symmetrical white spotting patterns are probably ancient and occur in breeds in Asia, Europe, the Americas, and perhaps also Africa. As with some other patterns, the symmetrical white patterns in European breeds can frequently be traced to Spain. White horses occur sporadically in a variety of breeds as distantly related as the Brabant and the Thoroughbred and may represent different origins of the white horse.

While many of the patterns of white hairs can be traced to a Spanish influence (in European breeds), it is difficult to go further back. The Spanish horse was a somewhat composite breed of many influences both from Europe and Africa, but exactly which influences brought which colors and patterns is a mystery. Many of the present variations in both color and pattern, especially in the New World, can be traced to the Spanish horse of A.D. 1200 to 1600.

Bibliography

Abeles, H. M.-S. 1979. A coat of many colors. *Equus* 17:30–38.

Adalsteinsson, S. 1974*a*. Color inheritance in farm animals and its application in selection. In *1st World Congress on Genetics in Relation to Animal Breeding*, pp. 29–37.

———. 1974*b*. Inheritance of the palomino color in Icelandic horses. *J. Hered.* 65:15–20.

———. 1978. Inheritance of yellow dun and blue dun in the Icelandic Toelter horse. *J. Hered.* 69:146–48.

Arabian Horse Registry of America (A.H.R.). 1970. *Identifying the Arabian*. Englewood, Colo.: A.H.R.

Berge, S. 1963. Heste fargenes Genetikk. *Tidsskr. Norske Landbr.* 70: 359–410.

Blunn, C., and C. Howell. 1936. The inheritance of white facial markings in Arab horses. *J. Hered.* 27:293–99.

Butage, R. 1974. Inheritance of coat colors in Belgian horses. *Vlaams Diergeneeskundig Tijdschrift* 43:464–86.

Castle, W. 1940*a*. *Mammalian genetics*. Cambridge.

———. 1940*b*. The genetics of coat color in horses. *J. Hered.* 31: 127–28.

———. 1942. The ABC of color inheritance in horses. *J. Hered.* 33: 23–25.

———. 1946. Genetics of the palomino horse. *J. Hered.* 37:35–38.

———. 1951. Dominant and recessive black in mammals. *J. Hered.* 42:48–50.

———. 1952. The eumelanine horse: Black or brown. *J. Hered.* 43:68.

———. 1954. Coat colour inheritance in horses and in other animals. *Genetics* 39:35–44.

———. 1961. The genetics of the claybank-dun horse. *J. Hered.* 52: 121–22.

———, and F. King. 1951. New evidence of the genetics of the palomino horse. *J. Hered.* 42:61–64.

———, and W. Singleton. 1960. Genetics of the brown horse. *J. Hered.* 51:127–31.

———, and F. Smith. 1953. Silver dapple, a unique color variety among Shetland ponies. *J. Hered.* 44:139–46.

Crew, F., and B. A. Smith. 1930. The genetics of the horse. *Biblthca genet.* 6:123–70.
Denhardt, R. M. 1975. *The horse of the Americas.* 2d ed. Norman: University of Oklahoma Press.
Dobie, J. F. 1952. *The mustangs.* Boston: Little, Brown.
Domanski, A., and R. Prawochenski. 1948. Dun coat colours in horses. *J. Hered.* 39:367–71.
Gallardo, R. C. 1936. *El libro del charro mexicano.* Mexico City.
Geurts, R. 1977. *Hair colour in the horse.* London: J. A. Allen and Co., Ltd.
Green, B. K. 1974. *The color of horses.* Flagstaff, Arizona: Northland Press.
Gremmel, F. 1939. Coat colors in horses. *J. Hered.* 30:437–45.
Haines, F. 1963. *Appaloosa, the spotted horse in art and history.* Austin: University of Texas Press.
Halversorn, J. 1977. The leopard that didn't get his spots. *Appaloosa News,* Dec., pp. 62–64.
Hatley, G. 1962. Crosses that will kill your color. *Appaloosa News,* vol. 19.
Hawkins, R. 1965. *The appaloosa: Breed characteristics.* Riverside, Calif.: Hawkins & Hubbell.
Jones, W. 1971. Appaloosa color inheritance. *Appaloosa News* 18 (1): 26–30.
———, and R. Bogart. 1971. *Genetics of the horse.* East Lansing, Michigan: Caballus Publications.
Klemola, V. 1933. The pied and splashed white patterns in horses and ponies. *J. Hered.* 24:65–69.
Marrero y Galindez, A. 1945. *Cromohipologia.* Buenos Aires: privately published.
McCann, L. 1916. Sorrel colour in horses. *J. Hered.* 7:370–72.
Miller, R. W. *Appaloosa coat color inheritance.* Moscow, Idaho: The Appaloosa Horse Club, Inc.
Norton, D. 1949. *The palomino horse.* Los Angeles: Bordon.
Odriozola, M. 1951. *A los colores del caballo.* Madrid: Publicaciones del Sindicapo Nacional de Ganaderia.
Pulos, W., and F. Hutt. 1969. Lethal dominant white in horses. *J. Hered.* 60:59–64.
Salisbury, G., and J. Britton. 1914*a*. The inheritance of equine coat color. I: The basic colors and patterns. *J. Hered.* 32:235–40.
———, and ———. 1914*b*. The inheritance of equine coat color. II: The dilutes. *J. Hered.* 32:255–60.
Searle, A. 1968. *Comparative genetics of coat colour in mammals.* London: Logos.
Singleton, R. 1969. The genetics of mammalian coat color. *J. Hered.* 60:25–26.

———, and Q. Bond, 1966. A allele necessary for dilute coat color in horses. *J. Hered.* 57:75–77.
Skorkowski, E. 1972. Colour and constitutional peculiarities go together (Darwin, 1859). *World Review of Animal Production* 8(4): 70–75.
Smith, A. 1925. A study of the inheritance of certain color characters in the Shorthorn breed. *J. Hered.* 16:73–84.
Sponenberg, D. P. 1982. The inheritance of leopard spotting in the Noriker horse. *J. Hered.* 73:357–59.
Sturtevant, A. 1912. A critical examination of recent studies in color inheritance in horses. *J. Hered.* 2:41–52.
Trommershausen-Smith, A. 1972. Inheritance of chin spot markings in horses. *J. Hered.* 63:100.
———. 1977. Lethal white foals in matings of overo spotted horses. *Theriogenology* 8:303–11.
———. 1978. Linkage of tobiano coat spotting and albumin markers in a pony family. *J. Hered.* 69:214–16.
———. 1979. Positive horse identification, part 3: Coat color genetics. *Equine Practice* 1:24–35.
———; Y. Suzuki; and C. Stormont. 1976. Use of blood typing to confirm principles of coat color genetics in horses. *J. Hered.* 67: 6–10.
VanVleck, L. D., and M. Davitt. 1977. Confirmation of a gene for dominant dilution of horse color. *J. Hered.* 68:280–82.
Wentworth, E. 1914. Color inheritance in the horse. *Z. indukt. Abstamm. Vererblehre* 2:10–17.
Wiersema, J. K. [1977?] *Het paard in Zijn Kleurenrijkoom.* The Hague: Zuidgroep B.V. Uitgevers.
Wilson, J. 1912. The inheritance of the dun coat colour in horses. *Proc. R. Ir. Acad.* 13:184–201.
Wright, S. 1917. Color inheritance in mammals. *J. Hered.* 8:560–64.

Index